Science

Compiled by
Marlene Peterson

Well-Educated Mother's Heart Learning Library
Libraries of Hope

Science

Compiled from:
The Boys and Girls Book of Science, by Charles Kingsley, London: Strahan and Co., (1881)

Early Chapters in Science, by Frances Awdry, London: John Murray, (1899).

The Chemistry of Creation, by Robert Ellis, London: W. Clowles & Sons, (1850).

Scientific Christian Thinking for Young People, by Howard Johnson, New York: George H. Doran, (1922).

Libraries of Hope, Inc.
Appomattox, Virginia 24522

Website www.librariesofhope.com
Email: librariesofhope@gmail.com

Printed in the United States of America

CONTENTS

THE BOYS' AND GIRLS' BOOK

OF SCIENCE

WITH NUMEROUS ILLUSTRATIONS

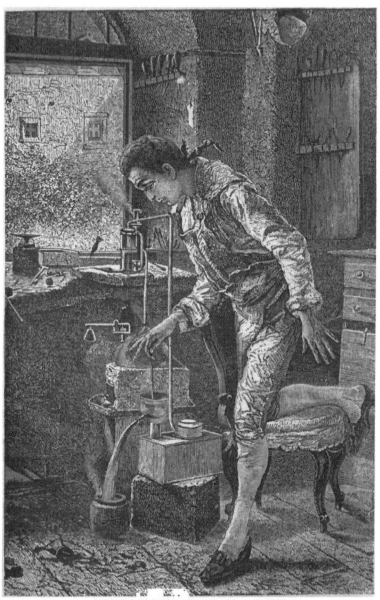

JAMES WATT INVENTING THE STEAM-ENGINE.

ADDRESS TO BOYS AND GIRLS.

M Y DEAR BOYS AND GIRLS,—When I was your age there were no such children's books as there are now. Those which we had were few and dull, and the pictures in them ugly and mean, while you have your choice of books without number, clever, amusing, and pretty, as well as really instructive, on subjects which were only talked of fifty years ago by a few learned men, and very little understood even by them. So if mere reading of books would make wise men, you ought to grow up much wiser than us old fellows. But mere reading of wise books will not make you wise men ; you must use for yourselves the tools with which books are made wise, and that is—your eyes and ears and common sense.

Now, among those very stupid old-fashioned boys' books was one which taught me that, and therefore I am more grateful to it than if it had been as full of wonderful pictures as all the natural history books you ever saw. Its name was "Evenings at Home," and in it was a story called "Eyes and No Eyes," a regular old-fashioned, prim, sententious story ; and it began thus :—

"Well, Robert, where have you been walking this afternoon?" said Mr. Andrews to one of his pupils at the close of a holiday.

Oh, Robert had been to Broom Heath and round by Camp Mount, and home through the meadows. But it was very dull. He hardly saw a single person. He had much rather have gone by the turnpike road.

Presently in comes Master William, the other pupil, dressed, I suppose, as wretched boys used to be dressed forty years ago, in a frill collar and skeleton monkey-jacket, and tight trousers buttoned over it and hardly coming down to his ankles, and low shoes which always came off in sticky ground ; and terribly dirty and wet he is ; but he never, he says, had such a pleasant walk in his life, and he has brought home his handkerchief (for boys had no pockets in those days much bigger than keyholes) full of curiosities.

He has got a piece of mistletoe, and wants to know what it is ; and he

3

has seen a woodpecker and a wheat-ear, and gathered strange flowers on the heath, and hunted a peewit because he thought its wing was broken, till, of course, it led him into a bog, and very wet he got. But he did not mind it, because he fell in with an old man cutting turf, who told him all about turf-cutting and gave him a dead adder. And then he went up a hill and saw a grand prospect, and wanted to go again and make out the geography of the country from Cary's old county maps, which were the only maps in those days. And then, because the hill was called Camp Mount, he looked for a Roman camp, and found one; and then he went down to the river and saw twenty things more, and so on, and so on, till he had brought home curiosities enough and thoughts enough to last him a week.

Whereon Mr. Andrews, who seems to have been a very sensible old gentleman, tells him all about his curiosities. And then it comes out—if you will believe it—that Master William has been over the very same ground as Master Robert, who saw nothing at all!

Whereon Mr. Andrews says, wisely enough, in his solemn, old-fashioned way :—

"So it is. One man walks through the world with his eyes open, another with his eyes shut; and upon this difference depends all the superiority of knowledge which one man acquires over another. I have known sailors who had been in all the quarters of the world, and could tell you nothing but the signs of the tippling-houses and the price and quality of the liquor. On the other hand, Franklin could not cross the Channel without making observations useful to mankind. While many a vacant thoughtless youth is whirled through Europe without gaining a single idea worth crossing the street for, the observing eye and inquiring mind find matter of improvement and delight in every ramble. You then, William, continue to use your eyes. And you, Robert, learn that eyes were given to you to use."

So said Mr. Andrews; and so I say to you. Therefore I beg all good boys and girls among you to think over this story, and settle in their own minds whether they will be Eyes or No Eyes; whether they will, as they grow up, look, and see for themselves what happens; or whether they will let other people look for them, or pretend to look, and dupe them, and lead them about—the blind leading the blind, till both fall into the ditch.

I say "good boys and girls;" not merely clever boys and girls, or prudent boys and girls; because using your eyes or not using them is a question of doing right, or doing wrong. God has given you eyes, and it is your duty to God to use them. If your parents tried to teach you your lessons in the most agreeable way, by beautiful picture-books, would it not be ungracious, ungrateful, and altogether naughty and wrong, to shut your eyes to those pictures, and refuse to learn? And is it not altogether naughty and wrong to refuse to learn from your Father in heaven, the Great God who made all things, when He offers to teach you all day long by the most beautiful and most wonderful of all picture-books, which is, simply all things which you can see, and hear, and touch, from the suns and stars above your heads, to the mosses and insects at your feet? It is your duty to learn His lessons, and it is your interest likewise. God's Book, which is the Universe, and the reading of God's Book, which is Science, can do you nothing but good, and teach you nothing but truth and wisdom. God did not put this wondrous world about your young souls to tempt or to mislead them. If you ask Him for a fish, He will not give you a serpent. If you ask Him for bread, He will not give you a stone.

So use your eyes and your intellect, your senses and your brains, and learn what God is trying to teach you continually by them. I do not mean that you must stop there, and learn nothing more : anything but that. There are things which neither your senses nor your brains can tell you; and they are not only more glorious, but actually more true and more real, than many things which you can see or touch. But you must begin at the beginning in order to end at the end ; and sow the seed if you wish to gather the fruit. God has ordained that you, and every child which comes into the world, should begin by learning something of the world about him by his senses and his brain ; and the better you learn what they can teach you, the more fit will you be to learn what they cannot teach you. The more you try now to understand *things*, the more you will be able hereafter to understand men, and That which is above men. You begin to find out that truly Divine mystery, that you had a mother on earth, simply by lying soft and warm upon her bosom ; and so (as our Lord told the Jews of old) it is by watching the common natural things around you, and considering the lilies of the field, how they grow, that you will begin at least to learn that far Diviner mystery—that you have a Father in heaven. And so you will be delivered (if you will) out of the tyranny of darkness, and distrust, and fear, into God's free kingdom of light, and faith, and love ; and will be safe from the venom of that tree which is more deadly than the fabled

5

Upas of the East. Who planted that tree I know not, it was planted so long ago; but surely it was none of God's planting, neither of the Son of God: yet it grows in all lands, and in all climes, and sends its hidden suckers far and wide—even (unless we be watchful) into your hearts and mine. And its name is the Tree of Unreason, whose roots are conceit and ignorance, and its juices folly and death. It drops its venom into the finest brains, and makes them call sense nonsense, and nonsense sense; fact fiction, and fiction fact. It drops its venom into the tenderest hearts, alas! and makes them call wrong right, and right wrong; love cruelty, and cruelty love. Some say that the axe is laid to the root of it just now, and that it is already tottering to its fall; while others say that it is growing stronger than ever, and ready to spread its upas-shade over the whole earth. For my part, I know not, save that all shall be as God wills. The tree has been cut down already, again and again, and yet has always thrown out fresh shoots, and dropped fresh poison from its boughs. But this at least I know, that any little child who will use the faculties which God has given him, may find an antidote to all its poison in the meanest herb beneath his feet.

There—you do not understand me, my boys and girls; and the best prayer I can offer for you is, perhaps, that you should never need to understand me; but if that sore need should come, and that poison should begin to spread its mist over your brains and hearts, then you will be proof against it, just in proportion as you have used the eyes and the common sense which God has given you, and have considered the lilies of the field, how they grow.

CHARLES KINGSLEY.

6

EARLY CHAPTERS
IN SCIENCE

A FIRST BOOK OF KNOWLEDGE
OF NATURAL HISTORY, BOTANY, PHYSIOLOGY
PHYSICS AND CHEMISTRY
FOR YOUNG PEOPLE

BY

Mrs. W. AWDRY

EDITED BY

W. F. BARRETT

PROFESSOR OF EXPERIMENTAL PHYSICS IN THE
ROYAL COLLEGE OF SCIENCE FOR IRELAND

WITH NUMEROUS ILLUSTRATIONS

PART I.

THE ANIMAL AND VEGETABLE KINGDOMS.

CHAPTER I.

CLASSIFICATION.

PROBABLY, when any one speaks of animals and animal life, we all think first of horses and cows, cats and dogs, and other conspicuous or useful animals, or perhaps of lions, tigers, and elephants.

But a little thought will remind us of many other varieties of living creatures, some of them differing so widely from these that they seem to have nothing in common but the fact of their all being alive.

Suppose we sit down quietly in a shady country garden in early summer, and make a list of the live creatures that come under our notice in the course of half an hour.

The first we see make us smile, for they are baby and the kitten, having a game of play; and next a flock of sheep passes bleating along the road, while the busy and important sheep-dog barks at the stragglers. But when they are gone by and all is quiet again, shyer creatures begin to venture in sight. What is that bonny ball of brown fur balancing on the branch of the larch tree, and nibbling something held in its paws? It is not every garden that is enlivened by squirrels with their pretty gambols, but I know one that is their constant and welcome haunt.

Now the whole air is musical with the voices of the birds; the thrush is never tired of his song, a cuckoo is proclaiming himself from the wood, numberless small birds are busily twittering and chattering over their young families, and here comes a swallow darting and turning after the flies. Ah yes, the flies; why, there are crowds and crowds of them dancing in the sunny air, and supplying food enough alike for the birds and the spiders. At midday the flies are but a slight annoyance; but if we linger in the garden till evening we know well that at five o'clock, or thereabouts, their places will be taken by swarms of midges, to the sorrow of any one who is sensitive to midge bites; and when the swallows go to bed, out will come the bats pursuing gnats and midges with their swift silent flight.

If we look attentively at the rough ground under the shrubbery we shall see that the fallen leaves and rubbish lying there are constantly in motion. What is moving them? On examining a little more closely we find beetles, snails, ants, and numberless other small creatures at work there. Turn over that log of wood, and many earwigs hurry off in all directions on important business, while the woodlice are trying to roll themselves up into pills, and a toad which had sheltered there, crawls leisurely into a cooler and darker retreat.

Our list may grow much longer yet. There are

butterflies at play in the sunshine, green aphis on the rose trees, bees buzzing about the flowers, while the ripple of the stream at the foot of the hill reminds us that its cool waters are full of fish. I do not suppose we shall see a garden snake or a lizard, as these are comparatively rare. But we want another creature that has not appeared yet, so let us walk round where the gardener is digging among the vegetables, and see what he turns up. Ah, here it is, a fine fat wriggling earthworm! Lay it on the hard path where we can have a good look at it. Is it not like a snake? They are both long, narrow, round creatures, without legs, which wriggle along upon the ground. But is a worm a snake? No; you shake your heads, you know better than that. Yet I am not so sure that you can say what is the great difference between them—a difference so important as to put them on opposite sides in the one great division that runs through all the animal creation; probably you have never looked inside either of them to see how they are made.

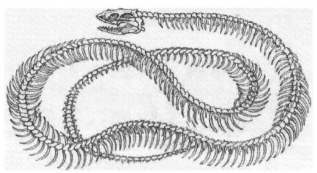

Skeleton of Snake.

Here is a picture of a snake's skeleton. If the skin and the flesh of a snake were all taken away, we should find these bones left—a head, and a long flexible backbone with numerous ribs coming out of it, bent round somewhat into the shape of the creature—a framework of a snake. Well, what about the worm? Ah, we are beginning to see the difference. A worm has no bones at all; if the body of a worm is cut open, its structures are found to be all quite soft; there is no hard, bony, framework. However alike they may outwardly appear, this is indeed a main distinction, and by means of it we may arrange all animals into two divisions—animals that have a bony skeleton on which the body is built up, and animals that have not.

Take the list of creatures in the garden, and see how they are divided by this test. Now that your attention is drawn to it, no doubt you can tell in a moment in which division most of them are to be placed. Certainly baby has bones; and the sheep, for we have seen mutton bones; and dogs and cats, for we have known of their bones being broken in traps; yes, and so have the squirrels, and, in fact, all the four-legged animals. What about the birds? Every one who has seen a fowl or any other bird on the dinner-table knows the look of its bones, and if birds are alike they must all have bones. Then, fish-bones are familiar enough to everybody, and some of us think fish hardly worth eating on account of them. Has the frog bones? Perhaps you are not so certain about this; but here is a picture of a frog's skeleton, which puts the matter beyond a doubt.

But they all seem to have bones; what is there left to go into the other class? Let us look back at our list, and see what else there is. Snails and slugs. Their bodies are soft enough. We may examine them without finding any trace of hard frame-work; and the flies, bees, and butterflies, if trodden on, will be crushed quite flat; there is nothing inside them hard enough to resist pressure. And the same may be said of the spiders and woodlice, as well as the worms. Most of the insects, indeed, have rather hard cases, but no skeleton inside. These are all *little* creatures; and

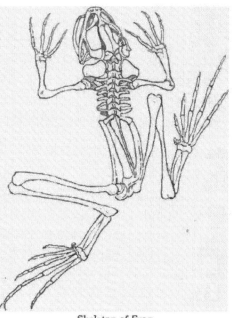

Skeleton of Frog.

though there are a few large ones in the sea, yet the great majority of the boneless creatures are comparatively small, for which, however, they amply make up by their countless numbers, both on land and in the water.

We need go no further than the two pictures of skeletons already given, to see that there is a good deal of difference in the number and shapes of bones in different creatures. Some creatures have legs, some have tails, some have both, and bones to support them are present or absent accordingly. Every animal, however, that has any bones at all, always has a backbone, formed of separate pieces or joints, more or less easily movable one upon another; and from this the whole great division gets its scientific name of *Vertebrate Animals*, or *Vertebrata*, from the Latin word *vertebra*, a joint, which is used of the joints of the backbone. The boneless creatures are called *Invertebrate Animals*, or *Invertebrata;* but we will put these on one side for the present, and, turning our attention to the Vertebrate Animals, try if we can find some other good test by which to divide them again into two sets.

Take this frog in your hand, and tell me what it feels like. It is not poisonous, there is nothing to be afraid of; yet it is an unpleasant, cold, slimy creature, and no one is very willing to handle it. But we do not mind touching other creatures. Only see how pussy is caressed. Ah, pussy is not cold and slimy; she is quite soft and warm. Warm and cold! Well, take this for the next division, and see which of the Vertebrate creatures are warm and which are cold. We can feel the dogs and the sheep, the cows and the horses, and find that they are all warm. The wild birds will not let us touch them, but in the poultry-yard, perhaps, we can get hold of a downy young chick newly hatched, and feel what a hot little thing it is. And when the hen chaffinch flies off her nest we can feel her eggs, which are quite warm from the heat of her body. To be sure, birds generally keep their eggs warm by sitting on them, so they must be all warm themselves.

Speaking of birds' eggs reminds us that we have yet one more division to make among the creatures with

13

warm blood: we must separate those that lay eggs from those that do not. The cows, horses, sheep, and many other animals are born alive, and nourished for some time by their mothers' milk. Creatures born and nourished in this way are called *Mammals*, or *Mammalia*, and all the warm-blooded animals known are either Mammals, or else belong to the *Birds*, which are, as we know, first produced as eggs, out of which the young birds are hatched under the influence of heat.

Now, let us in the same sort of way try and divide into two or three classes the cold and slippery Vertebrata, among which, besides frogs and toads, we must reckon the snakes and lizards, and all the fish. These creatures are all produced from eggs, which are generally deposited before hatching, although in a few instances the eggs are retained in the body of the mother until they are hatched, so that the young are born alive. We must therefore look for some other dividing line among them, and a very important one is afforded by their manner of breathing.

If we watch a live fish in a pan of water, we shall see a constant heaving movement of two openings set one on each side of the head, an apparent opening and closing of little doors. The little doors are the gill-covers, and if they were removed we should see behind them the gills, a series of delicate membranes so transparent that we can see the blood through them, making them look quite red. The gills are the breathing organs of fishes ; for fishes can only breathe air contained in water, and die when the gills become dry. But gills, which require to be always wet, would not at all suit the snakes and other creatures living on dry land, and they breathe by lungs, like the mammals and birds. The name for most cold-blooded Vertebrate animals breathing by lungs is *Reptiles*, while those that breathe by gills belong to the Class of *Fish.*

In which of these classes, then, are we to put the frogs and toads? Why, they can live on land and breathe dry air, so we take it for granted that as they have lungs they must be Reptiles. Ah, but think of their history! When frogs' eggs are hatched, frogs do not come out, but tadpoles, which swim about in water and breathe by gills like fishes; then, as they grow older, they gradually develop lungs, lose their gills, turn into frogs, and come out of the water. Gills first and lungs afterwards. Then they ought to belong to both classes. In fact, they are reckoned a separate Class, which stands between the other two, and is named *Amphibians*, or *Amphibia*, from two Greek words, meaning " life in both ways."

Thus we have arranged all the Vertebrate animals into five classes.

I. *Mammals.*—Creatures with warm blood, whose young are born alive,* and nourished by their mothers' milk.

II. *Birds.*—Creatures with warm blood, whose young are hatched out of eggs.

III. *Reptiles.*—Creatures with cold blood, breathing only by lungs.

IV. *Amphibians.*—Creatures with cold blood, breathing both by gills and by lungs.

---

* Except the Australian duckbill, and spiny anteaters, which lay eggs from which the young are hatched. They form the lowest order of mammals, showing, by their structure, affinity to the Reptiles.

V. *Fishes.*—Creatures with cold blood, breathing only by gills.*

It is clear that we might have taken quite different tests to form classes by; for instance, we might have reckoned together all the creatures that have four legs, so putting frogs and lizards into the same class as the mammals; or have made a test of the power of flying, which would include bats among the birds, while it left out ostriches. But the arrangement given above is that agreed upon by all naturalists, and the proof of a good arrangement is when the members placed in one class have many characters in common and do not depend upon one alone. So, if flying were the only test of a bird, the ostriches would have to be placed in some other class; but when we find that they are clothed with feathers as birds, and only birds, are: that they have no teeth, but a horny beak, a character belonging to birds alone: that they lay eggs, out of which the young are hatched, and moreover, that close examination of their structure shows that the wing-bones are present though very small; then we can have no hesitation in classing them as birds, though they do not fly. Bats, on the contrary, are covered with hair, like most of the mammals; they have soft snouts, and a set of sharp teeth; their young ones are born alive and suckled by the mother; so that, in spite of their power of flight, they cannot be included with the birds from whom they differ in so many important characters.

So again, if we looked only at its habit of swimming

* Except the mudfishes, found in rivers in South America, Africa, and Australia, which breathe also by lungs.

and living in water, we might easily suppose a whale to be a fish; but when on further examination we find that its blood is warm, that it breathes by lungs and not by gills, coming up to the surface of the water to take breath, and moreover that the young do not come out of eggs, but are born alive and fed with milk, we are convinced that its true place is among the mammals, in spite of much that is fishlike in its form and habits.

Besides the five classes of animals mentioned above, the soft, jelly-like Sea-squirts (*Tunicata*) are now generally classed with Vertebrates, on account of the correspondence of their development with that of the bony animals. The Lampreys and the Lancelet, formerly grouped with the Fish, are now considered worthy to rank as independent Classes.

Now for the Invertebrate Animals. We must get another worm, a slug and a snail, a wasp, a butterfly, a caterpillar, a spider, and a centipede.

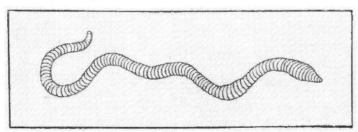

Worm.

If we were by the seaside, it would be well also to go down to the shore and search for a sea-anemone and a starfish, which may often be found thrown up on the beach after stormy weather; but if these are not to be had we must be content with their pictures. A shrimp,

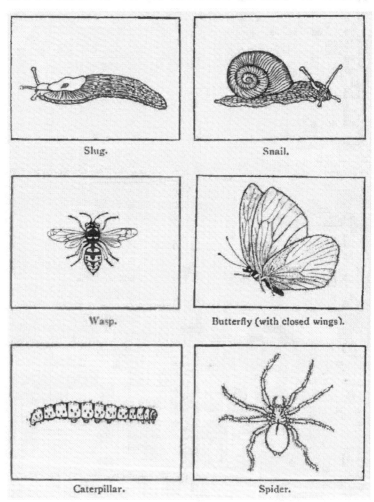

Slug.

Snail.

Wasp.

Butterfly (with closed wings).

Caterpillar.

Spider.

too, is wanted in our collection, and by the help of the fishmonger we can very likely get this anywhere.

First put the worm and slug side by side and notice

the difference in the way their bodies are made. The worm seems to be made in a number of different

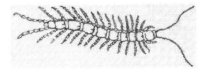

Centipede.

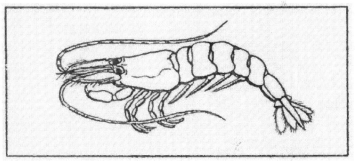

Shrimp.

Sea-anemone.

Starfish.

pieces or rings set one behind another all down the body from the head to the tail, while the slug shows no such divisions. Which of our other animals are made in the worm fashion, of pieces or segments joined one behind another? We pick out the caterpillar, the centipede, and the shrimp at once, but then comes a doubt. Well, never mind whether the segments are alike or not, provided they are behind each other. Upon this we add the wasp, butterfly, and spider to the set.

The snail's body is not in segments; it is more like that of the slug, only it carries on its back a snug house, into which it can withdraw on the approach of danger. Place it with the slug. Notice that both snail and slug walk upon a long, flat, muscular "foot."

Now there are the starfish and sea-anemone left. The starfish does indeed seem to be made of different pieces, but, instead of being set behind one another, they are arranged all round a centre in which is the creature's mouth ; and the anemone, though less divided up, is of the same general pattern. These two, therefore, must be placed together, apart from the other sets.

Here we have a few main distinctions among the Invertebrate creatures.

The first group, called *Mollusks*, or *Mollusca*, to which the slug and snail belong, have soft bodies enveloped in a muscular skin or mantle, and most of them have also a shelly covering, sometimes made like the snail's in one piece, sometimes like the oyster's in two. The "foot" is nearly always present in some form.

The animals whose bodies show segments one behind another may be divided into two groups, according to whether they have legs or not. Those that walk about upon jointed legs have been called *Arthropoda*, from two Greek words meaning "jointed legs;" while those that have no legs belong to the group of *Worms*. Among the *Worms*, too, are included a number of animals whose bodies are not divided into segments; flat-worms and thread-worms, for instance. These are separated from Mollusks by having no foot or mantle, though some of them live in shells (Lamp-shells).

Two more groups are formed of the creatures the parts of whose bodies are arranged in stars or circles round their own mouths. Those that have prickly or spiny skins, like the starfishes and sea-urchins, are called *Echinodermata* (thorny skins), and the smooth-skinned sea-anemones and jellyfish are known as *Zoophytes* (plant animals). Many Zoophytes, however, build up a hard framework to protect themselves, like the tiny coral animals; colonies of these, by their united labours, are able to form vast living reefs and islands. The best distinction between the echinoderms and the zoophytes will be seen if we look inside the animals.

In the starfish, as in all Vertebrates, Mollusks, and Arthropods, we shall find a distinct body-cavity between the inside of the body-wall and the outside of the food-canal (stomach, &c.). But the sea-anemone is merely a hollow bag with the inside lining thrown into folds; this hollow is the creature's stomach, and there is no body-cavity around it. Hence the sea-anemone and its allies are now called *Cœlenterata* ("hollow-stomachs") by most naturalists. In this group we can also include the *Sponges*, whose insides may be like a simple bag, or may become branched into a puzzling set of canals. But many naturalists nowadays prefer to reckon the sponges as a group by themselves.

There is yet another group, of which it is not easy to show you a specimen. Its members are creatures consisting of single *cells*, which are simple living bodies, generally so small that they cannot be seen except with the help of a microscope. One of the simplest is the little creature called an Amœba, which lives in ponds. These are the simplest forms of animal life, and are called *Protozoa.* In all the other groups of animals we have considered the body is built up of a great number of cells.

All of these groups are too large and general to be considered as "Classes" in the same sense as the five Classes of the Vertebrate animals. Each of them is to be compared with the whole group of Vertebrates, and each can be divided, like the Vertebrates, into several Classes; the Arthropoda, for instance, include at least four, and more likely six or seven, Classes. But, in truth, naturalists are not yet thoroughly agreed as to the proper classification of these myriads of little creatures, and we find different arrangements in the books of different writers. This list will, however, be a sufficient guide in the very short account that can here be given of the Invertebrate animals.

I. *Mollusks.*—Soft creatures, showing no rings or divisions along their bodies, generally provided with a "mantle" and a "foot," and often living in shells, such as shellfish, cuttlefish, slugs.

II. *Arthropoda.*—Creatures with bodies arranged in successive rings or segments, and with jointed legs, such as insects, centipedes, spiders, lobsters.

III. *Worms.*— Creatures with bodies arranged in successive rings, without legs ; also creatures without the ringed arrangement of the body and also without "mantle" or "foot." *

IV. *Echinodermata.*—Creatures with prickly skins, the parts of whose bodies are arranged round a central mouth, such as starfishes and sea-urchins.

V. *Cœlenterata.*—Creatures with smooth skins, and without body-cavities, usually with tentacles or soft arms arranged round a central mouth, such as jellyfish and sea-anemones.

VI. *Protozoa.*—Creatures consisting of simple cells.

In the diagram given on the next page the chief subdivisions of animal life are shown in a way that can be easily understood with a little care.

All living creatures on the earth, from the great whales and elephants to the tiny beings that can only be seen with a microscope, are included under one or another of the divisions given above; and we will now go on to learn something about the Orders and Families into which the principal Classes are divided. For though it would be an endless task to give even a very slight account of animals were each to be spoken of separately, yet by the help of this classification we may hope, even in a short space, to gain so much knowledge of the leading groups as to be able to refer to their proper places most of the animals we meet with, and to make it easier to remember anything we may learn about them in future. Here is a table which sets forth how all the

---

* Worms, in fact, form a very diverse group, whose members have hardly any characters in common, so that many naturalists do not regard them as one, but as several sub-kingdoms.

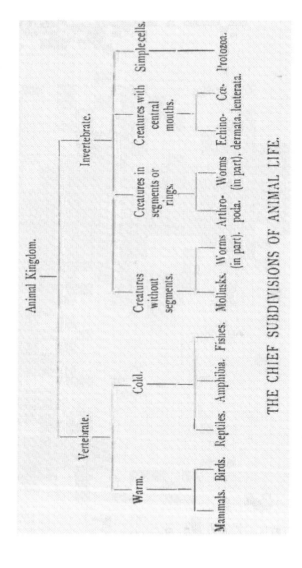

THE CHIEF SUBDIVISIONS OF ANIMAL LIFE.

Animal Kingdom.

Vertebrate.

Warm. Cold.

Mammals. Birds. Reptiles. Amphibia. Fishes.

Invertebrate.

Creatures without segments.

Mollusks. Worms (in part).

Creatures in segments or rings.

Arthro- Worms poda. (in part).

Creatures with central mouths.

Echino- Cœ- dermata. lenterata.

Simple cells.

Protozoa.

24

living creatures known upon the earth may be classified. The names in *italics* are those used by all naturalists, whatever language they may speak.

TABLE OF THE ANIMAL KINGDOM.

I. VERTEBRATE ANIMALS.

Class.	*Order.*	*Chief Groups.*
I. Mammals	i. Man * (*Bimana*).	
	ii. Apes, etc. (*Quadrumana*)	Monkeys and lemurs.
	iii. Bats (*Chiroptera*).	
	iv. Insect eaters (*Insectivora*)	Shrews, moles, hedgehogs.
	v. Beasts of prey (*Carnivora*)	Land.— Cat group, Dog group, Bear group. Water.—Walrus, sea-lions, seals.
	vi. Whales (*Cetacea*)	Including dolphins.
	vii. Manatee (*Sirenia*).	
	viii. Elephants (*Proboscidea*).	
	ix. Cony (*Hyracoidea*).	
	x. Hoofed animals (*Ungulata*)	Horse group, tapir, rhinoceros, swine, peccary, hippopotamus, oxen, sheep, goats, antelopes, giraffes, deer, chevrotains, camels, llamas.
	xi. Rodents (*Rodentia*)	Squirrels, rats, porcupines, hares and rabbits.

* The usual practice among zoologists is to class Man with Apes, Monkeys, and Lemurs in one order (*Primates*). For if the bodily structure of Man alone is considered, he is found to differ less from the higher apes than these do from the lemurs, or even from the lower monkeys.

Class.	Order.	Chief Groups.
	xii. "Toothless" animals (*Edentata*)	Sloths, anteaters, armadillos.
	xiii. Marsupials (*Marsupialia*)	Kangaroos, opossums, etc.
	xiv. Egg-laying mammals (*Monotremata*).	Duckbill, spiny anteaters.
II. Birds	i. Birds of prey (*Accipitres*)	Falcon group, owls.
	ii. Woodpeckers, etc. (*Picariæ*)	Climbing birds, parrots, etc. Wide-gaping birds, kingfishers, etc.
	iii. Perching birds (*Passeres*)	Crow group, thrush group, finch group, starling group, lyre bird group.
	iv. Pigeons (*Columbæ*)	
	v. Game birds (*Gallinæ*)	Fowl, turkey, pheasant, partridge.
	vi. Waders (*Grallæ*)	Plovers, cranes, rails, storks, snipes.
	vii. Water-birds (*Anseres*).	Ducks, geese, and swans, gulls and petrels, auks and penguins.
	viii. Wingless birds (*Struthiones*)	Ostrich, emu, cassowary.
III. Reptiles	i. Tortoises (*Chelonia*).	
	ii. Crocodiles (*Crocodilia*).	
	iii. Lizards (*Sauria*).	
	iv. Snakes (*Ophidia*).	
IV. Amphibians	(*Amphibia*)	Frogs, toads, newts.
V. Fishes	The orders of fishes are too difficult to be given in a small book like this.	

2. INVERTEBRATE ANIMALS.

Sub-Kingdom.	*Principal Classes.*	*Chief Groups.*
I. Mollusks (*Mollusca*)	Cuttle fish (*Cephalopoda*). Univalve shells (*Gasteropoda*). Bivalve shells (*Lamelli branchiata*).	
II. Creatures with jointed legs (*Arthropoda*)	Insects (*Insecta*). Centipedes and millipedes (*Myriopoda*). Spiders, etc. (*Arachnida*). Crabs and lobsters, etc. (*Crustacea*).	
III. Worms (*Vermes*).		
IV. Prickly skinned creatures (*Echinodermata*)		Starfishes, sea - urchins.
V. Zoophytes, etc. (*Cœlenterata*)		Jelly fish, sea - anemones, coral, and sponges.
VI. Simplest forms of life (*Protozoa*)		

We may arrange all these Classes, from the lowest and simplest forms of life to the highest and most complex, in a sort of genealogical tree, "with branches springing from different levels, each branch again bearing twigs, some of which rise higher than the base of the branch

27

above." At present our knowledge is very imperfect, but if we could make a perfect scheme of this sort, it would give not only an image of the wonderful unity

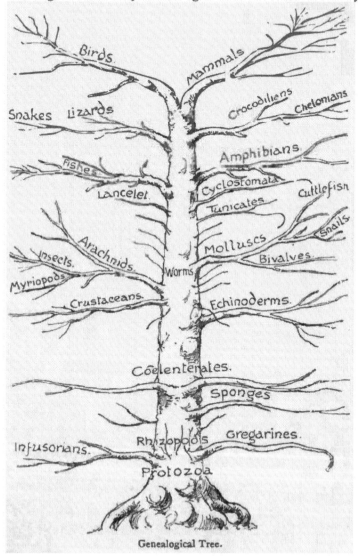

Genealogical Tree.

of the whole animal kingdom, but would also express the way in which naturalists believe the different forms of animal life are related to one another. Here is such a genealogical tree taken from Mr. J. A. Thomson's delightful book called " The Study of Animal Life." *

It must, however, be borne in mind that this picture is only an imaginary sketch, meant to be nothing more than a pleasing symbol, with many limitations and imperfections. It would be foolish to suppose any naturalist could assert that the various forms of animal life arose from one another exactly in the way shown in the picture. It will also be understood that this genealogical tree is not meant to express anything beyond the relationship of bodily structure in the animal kingdom.

* The small twigs from the main trunk represent the way in which different classes of *worms* may be supposed to be little off-shoots at different levels.

CHAPTER II.

Man.—Among the Mammalian creatures the first Order is that of human beings; but though this must be duly noticed, it is not our object to study them here. It is evident that in our bodily frames we closely resemble many other of the animal inhabitants of the world, and some account of this resemblance will be given later; but even the lowest races of men are immeasurably removed by their mental and spiritual capabilities from the lower animals.

Apes (*Quadrumana*).—We can easily recognize that next to man comes the order of animals most nearly resembling him, that of the Apes and Monkeys. The distinctive mark of the Order is in the formation of the hind paws, which always have five toes, one of them being placed opposite to the rest like our thumbs, so making a true *hand*, able to grasp firmly. Most of the apes and monkeys have also true hands on their fore-paws, but in some instances the thumb is very small, or altogether absent. Their skin is covered with hair, except on the face and the palms of the hands.

Monkeys are found both in the Old and New Worlds, but only in warm climates. When brought to our country they suffer much from cold, and are very liable to consumptive disease.

There is always something specially interesting to us in these "poor relations" of man, as they have been called; and no doubt the feeling that they are soulless caricatures of ourselves causes a sort of fascination in watching them, whether it is combined with horror at the fierce and hideous great apes, or with amusement at the grotesque impishness of the smaller monkeys. The likeness to man is not only in the outward form, but shows itself in the way in which many of them will weep with sorrow or distress, chuckle, and even smile, with amusement, redden with rage, or grow pale with fear; and in noticing these things we begin to have an uncomfortable sort of feeling that there is no very great difference between us. But if we imagine a monkey reading this book, or studying and classifying other creatures, we immediately see the absurdity of the comparison.

The largest of the apes is the great African Gorilla, which is between five and six feet in height when full grown, and is enormously strong in the arms and shoulders. The gorilla, like all the other apes, is without a tail, and when standing up has a horrible likeness to a very heavily-built and awkward man, with disproportionately long arms and hideous face. It cannot, however, stand upright without clinging to some support with its arms, and the knees are always somewhat bent, so that it does not really walk like a man. Its strength and ferocity are such that a full-grown male gorilla has never been captured alive, though the females and the

young have been taken. The gorilla lives in forests;
sometimes it climbs into the branches of the trees, and

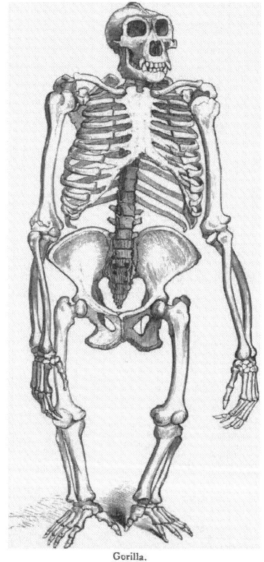

Gorilla.

when a band of negroes passes below it has been said to
reach down one powerful hind paw, and, seizing a man

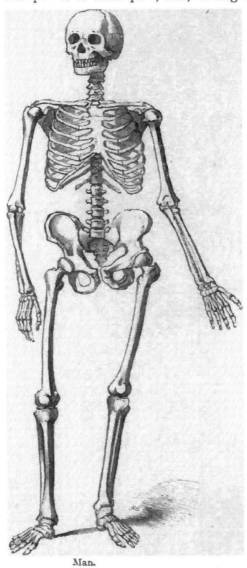

Man.

by the throat, to strangle him in its grasp. As it does not eat the man, this would appear to be done—if the story be true—from simple spite and malice.

The Chimpanzee, another African ape, is more gentle than the gorilla. Chimpanzees habitually live in trees, where they make nests among the branches.

Chimpanzee.

The rest of the apes are Asiatic, and consist of the Orang-outangs in Borneo and Sumatra—large creatures covered with reddish hair, living entirely among trees, whose branches they weave together into platforms to sit upon ; and of the Gibbons, which belong to Sumatra and Malacca, and are not quite such unpleasant-looking objects as the other apes. The Gibbons do not exceed three feet in height, and their bodies are slender and light, with small heads and very flexible necks, and extraordinarily long arms; some of them being able to put their whole hands flat on the ground without stooping. They are all very agile in their movements, springing from tree to tree almost as if they were flying. Their

Entellus Monkey.

disposition is shy and quiet, and they are capable of being tamed and trained in captivity.

Most of the members of the Order that have tails are called monkeys, and many different kinds of them are found in all the warmer parts of Asia, Africa, and America; but the only monkey inhabitant of Europe lives on the rock of Gibraltar, and is really a North African species that has crossed the Straits. It is known as the Barbary ape; but it is not a true ape, for, though its tail is reduced to a mere knob or projection, yet a tail it has. The monkeys have not the disproportionately long arms of the apes, and they generally go on all fours.

None of the family is better known than the Entellus, the monkey that haunts the villages of Northern India, and, being protected by the religion of the people, makes itself quite at home there. It varies from three to four feet in length, without reckoning the tail, which is often longer than the body, and it is covered with greyish fur, growing darker with age. Troops of these creatures live in the banyan groves, and make themselves a great nuisance by their cool familiarity and ingenious thefts, not only among the crops and fruits, but also in the shops and markets; yet it appears that even they are surpassed in impudence, grimaces, and mischief by the innumerable smaller monkeys of the African forests, who live a merry life chasing each other through the branches, chattering, screaming, playing practical jokes, and, when they get a chance, pulling out the tail-feathers of unfortunate parrots who come within reach of their mischievous fingers.

The Baboons of Africa have not the half-human look of other monkeys, but their heads are shaped more like that of a dog; a fact to which they owe their generic name of *Cynocephalus*. They hunt together in large troops, keeping strict discipline among themselves, and setting sentries to watch against the approach of enemies. The young are playful and impertinent; but they get soundly cuffed by the grave elders of the tribe if they misbehave, or make a noise when it is important that they should be quiet. They are large, strong animals, living among rocks, and making raids on cultivated grounds, where they commit great depredations.

The monkeys of America are entirely distinct, and there are no families of the tribe common to the Old

Baboon.

and New World. The expressive names Howlers, Spider monkeys, Squirrel monkeys, indicate some of their characteristics.

There is a tribe of small animals living in Madagascar and the neighbouring islands, sometimes called half-apes, but properly Lemurs. They have little monkey hands, and are therefore included in the same Order, but in appearance are more like foxes or cats, being rather pretty creatures, with very soft fur, and long, soft, round tails. They are nocturnal in their habits, sleeping through the day, but collecting together at night in large companies, and displaying great activity.

Lemur.

Bats (*Chiroptera*).—The next Order is that of the Bats, which are distinguished from all other Mammals

Bat, with spread wings.

by their wings and power of flight. Their bodies are

not unlike that of a mouse, but their forepaws have enormously long fingers, supporting the folds of large membranes, which connect the body, limbs, and tail, and form real wings. Bats are found in all parts of the

world, and vary in size, from the great Kalong, or Flying Fox of India, the spread of whose wings measures four feet across, to little creatures a couple of inches in length. Many of the large foreign bats are fruit eaters; but our English bats feed entirely on insects. People are often afraid of bats; but though their swift silent

Bones of the wing of a bat. (From Chambers's *Encyclopædia*.)

flight, and their appearance only at night, certainly give a feeling that there is a strange sort of mystery about them, yet there is nothing to be afraid of, and, indeed, if allowed to come freely into our rooms in warm weather, they often do good service by destroying gnats and other annoying insects. In the daytime they hook themselves up in some quiet dark corner, and wrapped up in their folded leathery wings are hardly to be recognized except by a practised eye.

Insect Eaters (*Insectivora*).—From the Bats we pass to another Order of creatures, who help to keep down our insect pests. This Order has not such clear distinguishing marks as the two already spoken of; but the creatures which compose it cannot be included in any other. The rats and mice seem very near them in some respects, but are clearly separated by their possession of the curious teeth which are the special mark of their own Order; while the smaller kinds of beasts of prey (Carnivora) differ in their bones and teeth.

So the Insect Eaters stand by themselves, consisting, in England, of the Mole, Shrew, and Hedgehog, and they have relations in all parts of the world except South America and Australia. The Mole lives, works, and hunts underground, constructing a wonderful dwelling, with tunnels leading into it in all directions, so that it is sure of a way of escape at the approach of an enemy. Worms and grubs are its food, and though, living in the dark, its eyes are small and little used, yet it is wonder-

Mole.

fully keen in hearing and smell, by which it detects and follows its prey.

In our country walks we are all familiar with the occasional sight of dead shrews lying about, though the cause of their death is unknown. Possibly their strong flavour prevents their being eaten by the creatures that kill them. They are not unlike mice in appearance, but the true mouse belongs to quite another Order of animals. The smallest known Mammal in the world is a tiny Italian Shrew, only an inch and a half long in the body, and with a tail of another inch.

The back of the Hedgehog is covered with prickly

spines, and it has the power of rolling itself up at the approach of danger, so as to present nothing but a ball of spines. It wanders at night in search of its food, trotting at a pretty rapid pace, and is not over particular as to what it takes. I have seen it eating a shrew, and it cannot be acquitted of the crime of feeding at times on young rabbits, poultry, etc. How a hedgehog walks up steep staircases is best known to itself; but when

Hedgehog.

kept in houses to eat the cockroaches, it may often be found in the most distant and apparently inaccessible parts of the building.

We might well be surprised at finding creatures so apparently unlike as Monkeys, Bats, and the Insect Eaters placed next to each other as the three first Orders of Mammals; but this is partly explained by the existence of a very curious animal in Malacca and the neighbouring islands, which seems to partake of the nature of all three. It has been known as the Flying Lemur, and used to be generally described as one of the Lemurs ; but naturalists now consider that its true place is among the *Insectivora*. Its limbs and tail are

41

connected by a large membrane, which spreads out and supports it while leaping from tree to tree, as if it were really flying; but it has not true wings, and cannot fly

Flying Lemur (*Colugo*).

upwards. It moves at night, and by day sleeps hanging by its hind feet from the branch of a tree, like a bat.

Beasts of Prey (*Carnivora*).—Now comes the vast order of beasts of prey; divided into the Land Carnivora, or eaters of flesh, and the Marine Carnivora, eaters of fish.

All the Land Carnivora are armed with sharp claws and teeth. Of front teeth, or *incisors*, they have six in the upper jaw and six in the lower. These are succeeded by a large strong pointed tooth on each side, corresponding to our eye-teeth, and properly called *canines*, which are the chief instruments by which the creature holds its prey. The back teeth are different in the different families. The six pair of incisor teeth are very characteristic of the order, for the only other animals which have them are the horse family, tapirs, and some swine, all of which are easily distinguished from the Carnivora by having hoofs and not claws; and also a few species of the Insectivora, which vary a good deal in the number of their front teeth.

Among the Land Carnivora are three main groups: the Cat group, including the families of the cats, the civets, and the hyenas; the Dog group, which has but a single family; and the Bear group, in which, besides the bear family, are reckoned the racoons, and the numerous family of the weasels.

The beasts of prey are powerful, active, energetic animals, with very acute sight and hearing. Unlike the quiet cattle which feed upon grass, they have to hunt their prey, and depend for a livelihood upon what they can catch and kill for themselves and their children. The young are born helpless, and often blind, so that they depend on the care of their parents, who gradually educate and train them to provide for themselves.

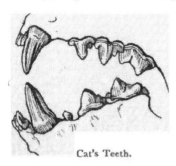

Cat's Teeth.

CATS.—The characteristics of the whole Cat family can be well seen in our domestic cats. Take pussy on your knee, and, if she will let you handle her, open her mouth and look at her teeth. Her front teeth are very tiny, but the canines are large and formidable, and you will see that the back teeth also have sharp points, not flattened crowns like our grinders, for as cats can only move their jaws up and down, and not at all sideways, they only chop and tear their food, and do not grind it. Their tongues are rough all over like a file.

Now, compare pussy's limbs with this picture of a lion's skeleton. If you pass your hand up her hind leg

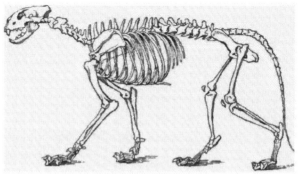

Skeleton of Lion.

you will find the knee-joint almost hidden in the loose folds of the skin of the body, and can then see that what looks like a knee turned backward, halfway down her leg, is really the heel, the beginning of a very long foot. The cats walk only on their toes, keeping this heel lifted well above the ground, which gives them a tread as light and springy as it is powerful. A cat always strikes with its feet, and, in the larger members of the family, there is immense strength in the feet and legs, for a lion will kill an ox with a blow of its paw, as a cat kills a mouse.

The toes are armed with sharp claws, which can be put out or retracted at will. This gives a double advantage; for when the claws are drawn back they are protected from getting blunted against the ground, and, at the same time, the soft cushions upon which pussy then treads, give the peculiar, silent, and stealthy step with which she steals upon her prey.

First of the great Cats stand the Lions. They are of a yellowish, tawny colour, with a tuft of hair at the end of the tail, and the head and shoulders of the male

Lion.

are covered with a thick mane, which gives him his majestic look. A full-grown lion measures about ten feet from the nose to the tip of the tail. His strength and swiftness are prodigious, and his roar strikes terror into the heart of every animal within hearing. He is not always anxious to attack. Dr. Livingstone writes, " When encountered in the day time, the lion stands a second or two, gazing, then turns slowly round, and walks as slowly away for a dozen paces, looking over his shoulder; then begins to trot, and when he thinks himself out of sight, bounds off like a greyhound." He had, however, another story to tell one day when he had attacked a lion, and shot at him, but missed his aim.

"He came with a tremendous roar, and Ferns" (Livingstone's horse) "whipped round like a top, and away at full speed. My horse is a fast one, and has run down the Gemsbok, one of the fleetest antelopes; but the way the lion ran him in was terrific. On came the lion, two strides to my one. I never saw anything like it, and never want to do so again. When he was within three strides of me, I gave a violent jerk on the near rein, and a savage dig at the same time with the off heel, armed with a desperate rowel, just in the nick of time, as the old mannikin bounded by me, grazing my right shoulder with his, and all but unhorsing me."

Livingstone then jumped off his horse and shot the lion.

Lions are found in all parts of Africa, and in the south-west corner of Asia; but the king of the Asiatic wild beasts is the royal Tiger.

The Tiger is of enormous size, sometimes reaching twelve feet in total length, extremely lithe, graceful and active in movement, and beautiful in colour, with black stripes or spots upon a yellow ground. He is a fearful enemy to cattle, and a tiger who has once tasted human flesh, generally becomes a confirmed man-eater, when no life is safe in his neighbourhood. "One Tiger, in 1887-8-9, killed, respectively, twenty-seven, thirty-four, forty-seven people." Tiger-hunting is generally undertaken by large bodies of armed men, mostly mounted on elephants.

The beautifully spotted Leopard, or Panther, occurs both in Africa and in Asia. It is smaller than the Lion or Tiger, but has an advantage over them in being able to climb trees.

In America the places of the lion and tiger are taken by the tawny-coloured Puma, and the Jaguar, which is spotted like a Leopard; and many smaller kinds of cats are found throughout America, Asia, and Africa, but Europe possesses only the Wild Cat, which, though nearly extinct in Great Britain, is still found in many parts of the Continent, and several varieties of Lynxes.

CIVETS.—Next to the true Cats comes the Family of Civets, of which we hear comparatively little, as none of them belong to this country. They vary considerably in size; the largest of them being about fifty inches in total length, the smallest not above twenty. In general appearance they are not unlike cats, but have longer and more pointed noses. They walk on their toes, but keep the heel much nearer the ground than cats, and their claws can only be partially drawn back, not entirely concealed. Their home is chiefly in Africa and Asia, but one species, the Genette, lives in Southern Europe.

HYENAS.—The Hyena may be considered to stand half way between the Cats and the Dogs. It is not unlike a hideous, degraded-looking dog, with a long, blunt nose and bushy tail. It has immensely strong teeth and jaws, which can crack even large bones; but it is a cowardly beast, and rarely kills food for itself, preferring to follow the great beasts of prey, and eat what they leave, or haunting burial-grounds, where it prowls at night in search of the dead bodies, uttering cries like

Hyena.

a horrible discordant laugh. It belongs to Africa and parts of Asia.

Dogs.—The Dog Family includes Dogs, Wolves, Foxes, and Jackals. Their claws cannot be drawn back at all; and though they walk on their toes, like the cats, yet they have not the same power in the paws; and when they attack, it is always with their teeth. The teeth also vary from those of the cats; for instead of having all sharp cutting teeth, the two hindermost teeth of dogs, above and below on each side, are grinders, with flattened crowns.

There are no animals that vary so largely as domestic dogs. They are of many colours, and of all sizes, from the huge mastiff to the tiny toy dogs petted by ladies. We have swift dogs, like the greyhound; sporting dogs, like the pointer and setter; fighting dogs, like the bull-dog; hunting dogs, like the foxhound or the terrier;

wise, responsible dogs, like the collies. No doubt this great variety is largely produced through careful breeding and training by men, but it remains doubtful whether they really all belong to the same species or not. Dogs are so constantly our companions that we know more of their minds, their intelligence, sympathy, even of their conscience, than is the case with any other animals. Who does not know how ashamed of himself a dog will look when he knows he has done wrong, how happy he will be when his master is pleased with him, how quickly he will learn what he is wanted to do, or how quiet and gentle he will be when he sees his friends sad and depressed?

In the Arctic regions the Eskimos are entirely dependent on their dogs to draw their sledges over the ice and snow when they move from place to place, and the dogs are also employed in hunting Bears and Seals, but they are only half-tamed, and are often very savage and unmanageable.

What is known as the Cape hunting dog, though it belongs to the same group, is not one of the true dogs ; but there are real wild dogs in India, and one in Australia, called the Dingo. Wild dogs live in burrows, caves, or hollow trees. It is a curious thing that wild dogs, and even domestic dogs that have run wild, whine or howl like wolves, but seldom or never bark ; it almost seems as if the barking of a civilized dog is like an attempt to imitate man in speaking.

FOXES.—Foxes have more pointed snouts, shorter legs, and bushier tails than dogs, and also the pupils of their eyes narrow into slits in bright light, like those of a cat. Their depredations among poultry and other small animals are well known.

Fox.

JACKALS.—The Jackal of Asia and Africa is not very unlike a yellow fox. It will follow larger beasts of prey, and eat what they leave, or will kill sickly or wounded animals for itself. Jackals hunt by night in packs, and their wailing cry as they sweep by an encampment is one of the most weird and melancholy of sounds.

WOLVES.—In the Wolf we come to a far more formidable animal. It is about the size of a large shepherd's dog, to which it bears a good deal of resemblance. It will kill and devour almost anything. Horses and oxen are pulled down by wolves; and in winter when they hunt in large packs, even a troop of armed men cannot always defend their lives from these savage beasts, unless they can reach some shelter. Wolves used formerly to live in the British Islands, and it is less than two hundred years since the last was killed in Scotland, while in Ireland they lingered also into the last century.

Wolf.

BEARS.—From the Dog group we pass to the Bear group. Bears have larger front teeth than cats or dogs, and grinders behind their canine teeth. They walk on the whole sole of the foot, which gives them a slow, heavy tread, very different from that of the light creatures that run on their toes. They have long sharp claws, which cannot be drawn back; and when attacked will rear themselves up on their hind legs, and strike terrible blows with their claws. Or they will try to seize an adversary in their front paws, and squeeze him to death. Most of the Bears are harmless enough towards men, unless attacked first or pressed by hunger. This, however, is not the case with the great Grizzly Bear of North America, nor with the Polar Bear of the Arctic Seas, both of which are very dangerous creatures, and most formidable foes to the hunter. Bears feed on

Bear.

mixed food, and most of them appear to be quite as partial to vegetable as to animal food, and particularly fond of sweet things. Indeed, none of the flesh-eaters object to an occasional change of diet, and even a lion is said to enjoy water-melons.

Weasel.

WEASELS.—Of Bears we happily see none in England except in menageries, but the same cannot be said of the large Family of the Weasels. Long, slender, and lithe in body, very short in the limbs, very sharp in the teeth, and very fierce in disposition, are most of its members. Weasels, Stoats, Martens, Polecats, Ferrets, have a strong family resemblance to each other, and are all terrible enemies to poultry and other small creatures. Ferrets are frequently used in hunting rabbits and rats, as they will go into the animals' burrows and turn them out. A larger cousin of theirs is the Badger, which more nearly resembles some of the foreign members of the family. Lastly we come to the Otter, which burrows in the banks of rivers, and lives upon fish,

Otter.

being itself equally at home in water or on land. The Sea-Otter of the North Pacific varies from other members of the Order in having only four lower incisors, and in any case, with its aquatic habits and its webbed feet, it forms a natural link to the next Family.

MARINE CARNIVORA.—The sea, as well as the land, has its beasts of prey, feeding upon the fish. These are the Walrus, the Sea Lions, and the Seals. The limbs of all these creatures scarcely resemble legs at all, but are modified into mere flappers, by which they move nimbly and easily in the water, but awkwardly on shore.

The Walrus is an enormous animal, from twelve to fifteen feet in length, and its great tusks, which are really the upper canine teeth immensely developed, give it an extraordinary and formidable appearance, which is indeed justified by its fury in fight when attacked. Walruses congregate in vast herds along the shores of

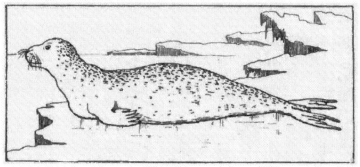

Seal.

the Arctic Seas, but upon any alarm they at once make for the water, where they feel safe.

Seals are remarkably intelligent creatures, easily tamed and very affectionate, and in captivity they readily learn to obey the voice of their keepers and to play many tricks. They are much attracted by music, and are well known to follow boats in which a musical instrument is played. The valuable seal-skin fur is not the skin of the true seals, but of a species of the sea lions, or eared seals, which are often much larger animals. The true seals do not exceed five feet in length, while some of the eared seals reach as much as ten feet.

All the Marine Carnivora come ashore for the birth

and nursing of their young, and by this power of coming ashore are completely marked off from the next Order of the Mammalia.

Whales (*Cetacea*).—Whales belong entirely to the water, and by ignorant people are often supposed to be fish; but though rather fish-like in form and without limbs, yet they are true Mammals, warm-blooded and suckling their young. To the Whale family belong the largest animals known, some of them exceeding even a

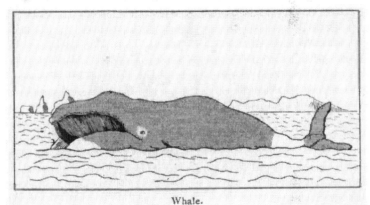

Whale.

hundred feet in length. They have to come to the surface from time to time to breathe, and if unable to do so would be drowned; but the intervals for which they can remain below vary in different species, from five minutes to an hour. Some of the whales have teeth, but other kinds have no teeth, their mouths being furnished with a series of plates, of what is commonly called whalebone, but more properly *baleen*. These plates are solid where they join the palate, but split into fringes at their loose edges, and serve as strainers to keep in the small shellfish and other creatures that enter the whale's mouth with the sea water. Huge as the whales are, they have very tiny throat passages, not more than two inches across. Hence they live principally upon the smaller kinds of crustaceans, mollusks, and fish.

The families of Dolphins and Porpoises, however, which belong to the same Order as the Whales, have not this difficulty, and make great havoc among fish.

Sirenia.—We will only notice in passing another small Order of strange sea beasts, to which belongs the Manatee, an ugly and ungainly animal, which, however, is supposed, from its habit of lifting itself upright in the water, and carrying its young in its arm, to have given rise to the legends about mermaids.

The animals of this Order cannot be included with the Marine Carnivora, which are all furnished with powerful canine teeth to hold their prey, while the Sirenia have no canines at all, and feed only on seaweeds. On the other hand, they are quite distinct from the Order of Whales, for the Cetacea have smooth bodies, and their nostrils or " blow-holes," as they are called, set right on the top of their heads ; while the Sirenia have hairy bodies, and their nostrils on the end of their snout. They must, therefore, be set in an Order by themselves.

CHAPTER III.

WE must now leave the sea, and return to the land, where we are approaching the vast Order of the Hoofed Animals; but a word must first be said about two Families which are by some naturalists considered to belong to it, while by others they are separated into distinct Orders, as not having true hoofs.

Elephants (*Proboscidea*).—The first of these are the Elephants, the largest of land animals, natives of Africa and India. The Elephant is a great heavy creature, usually about eight feet in height when full grown, but occasionally reaching ten or eleven feet, and of a dark grey colour. His great tusks, so much sought after by the ivory hunters, are not, like those of the Walrus, canine-teeth, but are the upper front teeth, or incisors, developed to a very great length; and he has also the advantage of growing fresh grinding teeth when the old ones are worn out. But the most wonderful thing about the Elephant is its trunk, which is really an immense lengthening of its nose. The trunk can be lengthened or contracted, waved from side to side, or curled round to carry food and water to the elephant's mouth, and its sensitive tip can pick up any small object from the ground. The Elephant has been humorously and happily described as "a square animal, with a leg at each corner, and a tail at both ends;" but few tails could be so useful as the trunk. Elephants are wonderfully wise and teachable beasts, and can be employed in many ways. Not only does their great strength enable them to carry all sorts of heavy burdens, but they will also learn to pile up stacks of logs, and even to lay courses of masonry in building, while it is well known that the keeper can safely trust his child to the care of one of these faithful attendants.

Coney Family (*Hyracoidea*).— The other Family spoken of above contains the little creature which in the Bible is called the Coney and a few related animals

Coney.

living in Africa. The group seems very small to have a Family and an Order to itself; but while the skeleton shows resemblances to those of several other animals, it cannot be exactly classed with any of them. Coneys are small thickset animals with short legs and ears. One group (including the animal mentioned in Psalms and Proverbs) live in colonies among rocks. Another, inhabiting South and West Africa, climb trees.

Hoofed Animals (*Ungulata*).—The Order of the Hoofed Animals must be divided into groups somewhat in the same way as the Order of the Beasts of Prey ; and the main distinction is made by the number of the toes. The first or *odd-toed* group includes the Horses and Asses, the Tapir and the Rhinoceros Families in which the middle toe is the longest ; in the Horse family this toe alone is developed and bears the hoof. All the rest of the hoofed animals belong to the immense group of the *even-toed* or cloven hoofs, which must be further subdivided when we come to it. All the animals of the Order feed on vegetables, except the Swine Family which eat everything they can get.

HORSE FAMILY.—The Horses and Asses, the creatures with a single hoof on each foot, are a good deal like each other, and form only one Family, but the Horses have horny places, or warts, on the inner side of each leg, and tails all covered with long hair, while the Asses have the warts only on the fore-legs, and hair only on the end of their tails.

Their teeth are not all close together, but there is a considerable space left between the front and back teeth, which affords room for the bit by which they can be controlled and guided.

The Horse, like the dog, is known to us in a great variety of breeds, but these seem to be due to careful continued selection, and not to any real difference of species. It is a highly intelligent and affectionate animal, and its good memory for a way once travelled has often been the saving of a rider who has lost his way; but horses are very nervous, and, if frightened or distrustful, will often appear ill-tempered or unmanageable when in truth they chiefly need to be soothed or reassured.

Our hardy, strong little Donkeys are not generally apt to be very swift in their movements, but most of the wild Asses are remarkable for swiftness and wariness. Wild Asses of different species are found both in Asia and Africa, and Africa is also the home of those striking and beautiful creatures, the Quagga and the Zebra, both of whom belong to the Asses. The domestic Ass has generally a single dark stripe across its shoulders, perhaps marking its relationship to the finely-striped Zebra.

TAPIRS.—Tapirs, which form the second division of

Tapir.

the Hoofed Animals, are creatures about the size of a donkey with very thick hides, and short trunks; they are fond of water and swim well. They are not very clean feeders, but live chiefly on vegetables, and never attack men unless hard pressed by hunters. Their home is in Central and South America, and in the Malayan peninsula and islands.

RHINOCEROS.—Perhaps there is hardly a more hideous beast living than the Rhinoceros. Its large, heavy body, sometimes reaching 12 ft. in length, and 5 ft. 10 ins. in height, and thick, tough skin, which is much prized

Rhinoceros.

by African and Indian natives for making shields, give it somewhat the appearance of a huge pig ; but, unlike a pig, it carries either one or two horns, not on the top of its head, but set behind one another along its snout. In the largest African Rhinoceros the front horn varies in height from 2 ft. 6 ins. to 4 ft., while the hinder one is only about 12 to 15 inches. In spite of its unwieldy appearance the Rhinoceros is a very swift runner, severely trying the powers even of good horses, and this combined with its uncertain temper and habit of making attacks without waiting for provocation, make it a very dangerous beast. Even the Elephant has a terror of the Rhinoceros' powerful horn. It is fond of wallowing in mud, partly perhaps as a protection against the insects by which it is constantly infested.

As the Families of the Swine and the Hippopotamus differ considerably from the other two-toed families, and in their heavy build and thick hides approach the character of the Rhinoceros and Tapir, and even of the Elephant, it is natural to take them first for consideration among the cloven-footed animals.

SWINE.—The Pigs or Hogs have round snouts, cut off abruptly at the end, and capable of being moved about a good deal, with which they root about, ploughing up the ground as they seek for food with their very keen scent. In their wild state they are neither stupid nor specially dirty, and if left to themselves they soon run wild and get back to the fierceness of their natural condition. It is not so very long since wild boars became extinct in the British Isles, and they are still found in most parts of Europe, Southern Asia, and Northern Africa. A large wild boar is a very powerful animal; armed with strong, sharp tusks, and being very swift in his movements, he will charge in a most dangerous manner.

In India the animals, which are fond of being in thick cover, often take up their abode in the standing crops, where they do great damage; they are hunted on horseback with spears. They grow to a great size, and a male has been measured as much as five feet nine inches in length; the females being smaller and having smaller tusks.

The hogs are replaced in America by smaller animals of the Swine Family called Peccaries, which do not exceed three feet in length. They look harmless enough, but are fierce little creatures, running in large troops; and being absolutely fearless, and armed with small, scarcely-seen tusks, as sharp as lancets, are dangerous to encounter.

HIPPOPOTAMUS. — The Hippopotamus, or River Horse, is found only in Africa. This enormous creature,

Hippopotamus.

though not exceeding about five feet in height, is frequently eleven or twelve feet long, and it opens its huge mouth with a width of gape unapproached by any other animal. It is of a heavy, unwieldy-looking build. The hippopotamus was first brought to England in 1850, and when full grown reached a weight of four tons.

These animals live by day almost entirely in the water,

where they may be seen together in large numbers, and where they can remain below for a considerable time between their breathing intervals; but at night they often come ashore, trampling and devastating the crops on which they feed. They are for the most part quiet in disposition when unprovoked; but a bull hippopotamus in a fury is an enemy not to be trifled with, and will easily crunch up a boat in its huge jaws, while the extreme thickness and toughness of its skin make it very difficult to kill.

RUMINANTS.—All the other families of the cloven hoofed animals, after the Pigs and Hippopotamus, are characterized by a curious habit. Any one who has watched a Cow knows that after feeding she will lie down in a quiet spot and enjoy the food over again. From the peculiar structure of the stomach she is able to bring the food again into her mouth, where it is leisurely chewed, and prepared for digesting. This action is called "chewing the cud," or *ruminating*, and the animals that perform it are called *Ruminants*.

The Ruminants include all cattle, sheep, and goats, all antelopes and deer, and the camels and llamas. Many of them have horns, which are always set side by side on the top of the head, not along the nose like those of the Rhinoceros. But there is an important distinction between their horns. The cattle, sheep, goats, and antelopes have hollow horns, covering a hard bony core, and they remain through life, gradually growing larger; but the horns or antlers of the deer family are solid, not hollow, and they fall off and are renewed every year.

OXEN.—No creature that lives is more valuable and useful to man than the common ox. In this country we value them chiefly, while living, for the sake of the milk of the cows; but elsewhere their strength is constantly used for labour, both in ploughing and in drawing waggons. Indeed in South Africa the principal means of travelling is in waggons drawn by teams of oxen, which are less liable than horses to the attacks of the terrible

Indian Ox or Zebu.

tsetse fly of that country. Then their flesh is one of our most valuable foods, and almost every part of the body can be turned to useful account.

Cows are so familiar to us that we can closely observe their ways, and it is very curious to see the order which they maintain among themselves. The leading cow of the herd is supreme in dignity; none of the younger animals will presume to enter or leave the pasture before her, and so tenacious is she of her position that it is said that when a leading cow in Switzerland was deprived of the deep-toned bell hanging round her neck which gave the signal to the rest, she refused her food and pined away, and though the bell was restored, it was too late to save her life.

In the Ox Family must be reckoned several kinds of Asiatic oxen, including the large Indian cattle with a

Bison.

hump on the back, the Yak, or long-haired ox of Tibet, the Bisons, and Buffaloes. These two last names are often mixed and applied to the same animals; but properly speaking, the Bisons include only the Aurochs, a nearly extinct animal of the forests of South-eastern Europe, and the North American Bisons. These last have thick shaggy manes and beards, and are large animals, the males being about six feet high at the shoulder. They move about in herds, which used formerly to be often vast in numbers; but they are now much reduced by hunters, who valued them for their skins and their excellent beef.

Buffaloes of different kinds are found in South Europe, in Africa, and in India, and are among the largest of the Ox tribe, with very large horns. Herds of tame buffaloes are kept like oxen; but wild ones are often dangerous, and have a special enmity against tigers, which they attack and kill. All the buffaloes are very fond of lying in mud and water, sometimes showing only their noses and eyes above the surface, and to this position they retire to chew the cud.

With the little Musk Ox of the extreme north of America we seem to be passing from the oxen to the sheep, as its appearance is not very unlike that of a ram covered with long hair, which hangs nearly to the ground, concealing its limbs. The name is derived from its strong musky odour.

SHEEP.—The Sheep Family rival the oxen in their usefulness to man, and have been so long domesticated that it is impossible to say from what country our breeds of tame sheep first came. We read in the very beginning of history that Abel was a keeper of sheep, and wherever men have migrated they have taken with them this docile animal, valuable equally for its flesh and its wool. The breeds vary greatly, especially in the presence or absence of horns, in one or both sexes, and also in activity and spirit, mountain sheep being generally remarkably agile.

Of the wild sheep the largest number belong to Asia;

but the Moufflon lives in Corsica and Sardinia ; North Africa has a large and handsome species in the Barbary sheep, and America in the Big Horn of the Rocky Mountains. All the wild sheep have horns, which in some kinds reach a very large size.

GOATS.—Closely related to the sheep, and really difficult to distinguish from them in some species, are the Goats, all of whom are horned and bearded. They belong exclusively to Europe, North Africa, and Asia. The flesh of the goat is not so highly esteemed as that

Mountain Goat or Ibex.

of the sheep, but their milk is more used, and the hair of the Angora and Cashmere goats is as valuable as the sheep's wool.

ANTELOPES.—The principal home of the great Antelope Family is in Africa, where vast herds of these creatures range in endless variety of species, varying in size from the Grand Eland, which rivals the ox in dimensions, to the tiny and graceful Gazelles, barely two feet high. We read in African travels of Steinboks, Springboks, Bushboks, Gemsboks, Koodoos, Hartebeests, and many others, all included among antelopes. Some are nearer the Ox group, others more like goats, while

Gazelle.

many have a strong resemblance to deer, from which, however, they are clearly distinguished by their permanent hollow horns. Most of them are light, active, and graceful in their movements, and the little Springbok, which is about thirty inches in height, frequently leaps into the air to a height of from seven or eight to as much as twelve feet. But the most extraordinary of the African antelopes is the Gnu, or Wildebeest. It has a head and shoulders not unlike a bull, while its hind quarters and tail are more like those of a pony, which animal it also

Gnu.

resembles in its manner of wheeling, prancing, kicking, and snorting.

Several species of Antelopes are found in Asia, one in California, and one in Europe, the pretty little Chamois of the Alps.

GIRAFFES.—Next to the Antelopes, and intermediate between them and the Deer, is placed the Giraffe, which, instead of true horns, has only two short appendages on the head, and a bony lump between the eyes, all of which are entirely covered by the skin.

There is something very attractive about this quaint, long-legged, long-necked creature, which, with its gentle eyes and awkward, angular movements, so vividly suggests a Noah's ark animal of wood and leather as to give rise to a constant wish to feel it and make sure that it is real. The great height, some sixteen or eighteen feet, from which it looks down on us should, indeed, inspire respect; but then it is all the more comical when straddling its legs apart in the difficult endeavour to get its head to the ground. In its African home it feeds habitually on the leaves of trees, selecting and plucking them daintily with its long tongue. It is a gentle, timid, and affectionate animal, and has rarely been heard to make a sound.

DEER.—The Deer Family are distinguished from all others by the remarkable history of their antlers. These ornaments belong to the male animal only, except with the reindeer, whose female is also horned. The horns vary greatly in size and in the number and kind of branches borne by different species and at different ages,

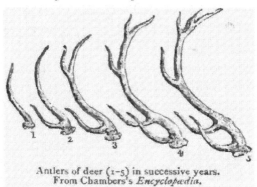

Antlers of deer (1-5) in successive years.
From Chambers's *Encyclopædia.*

but all alike are shed every year. In each year of a stag's growth its antlers grow more and more branched. The Red Deer, which are still found wild in parts of the British Islands, lose their horns in the spring between February and May. In a few days, however, they begin

71

sprouting again, and all the time they are growing they are covered with a soft furry skin, called " velvet," which is hot to the touch from the rapid coursing of the blood in it. When the growth of the horn is complete for the year, this skin or " velvet " dries up, and is gradually worn off by the animal rubbing its head against the trees until only the hard solid horn of the antler remains. In its first year the young stag grows only a simple spike with one point, but year by year the new horns increase in size and complexity until the full-grown stag over six years old, carries magnificent antlers with points varying from sixteen up to even as many as sixty-six. But the whole of this growth is completed in a very short time—about ten weeks. By the end of August the horns are cleared of " velvet," and the fully-armed stag, who remained in quiet and retirement while his weapons were growing, comes out prepared to fight the world, and at this time of year, is very quarrelsome and dangerous. The stags fight each other for the possession of the does, and not infrequently kill each other. The fawns, which are born in May and June, are brightly spotted with white in the summer, and gradually assume the red colour of the full-grown animal.

In those kinds of deer which have very small or simple horns, the canine teeth are developed into small tusks as if to balance the want of them, and there are one or two species altogether hornless, such as the Musk Deer, and the Chinese Water-deer, in which the tusks become of considerable size.

Deer belong chiefly to Europe and Asia; there are a few in America, but they are unknown in Africa south of the Sahara, where their place seems to be taken by the antelopes. The red deer and the roebuck are natives of Britain, though now driven into remote districts, but the fallow or spotted deer, which are most frequently seen in our parks, are imported from other countries.

The largest of the deer is the Elk, or the Moose of North America, which sometimes stands eight feet high

Head of Reindeer.

at the shoulder—as high as a fair-sized elephant. Its horns are "palmated;" that is, their branches are connected together by sheets of bony tissue, making the whole pair of antlers appear like a huge basin, and adding immensely to the weight, which is so great that one wonders how the animal can carry such a burden. Nevertheless, the Moose is a swift and powerful runner, and a good swimmer. An extinct gigantic deer, the so-called " Irish Elk," had still larger antlers than the Moose.

The most useful to man of the Deer Family, is the Reindeer of the Arctic Regions, herds of which constitute the wealth of the Laplanders. This is a powerful and enduring animal, and is used both for carrying riders and baggage, and drawing sledges. It is peculiarly fitted for travelling over snow, as the two sides of the hoof part widely when pressed upon the ground, and spread so much as to give the same sort of help as a snow-shoe. Also one of its horns has generally a branch widely expanded, and standing straight forward just over its brow (see picture), which serves as a snow-plough to shovel aside the snow, under which its food, consisting in winter principally of a dry sort of lichen, is concealed.

CHEVROTAINS.—After the Deer come a family known as the Chevrotains, which are wee, pretty Deerlets, about as big as a rabbit or hare, with large dark eyes and a gentle and confiding expression. They differ from the true deer in having no horns, but tusks large enough to show outside when the mouth is shut ; and also the bones of the feet approach in some respects more nearly to those of the swine. These graceful little creatures belong to Southern Asia and West Africa.

The whole of the Ruminating animals hitherto spoken

of are distinguished by having front teeth only in the lower jaw, the upper front teeth being replaced by a horny sort of pad against which the lower teeth bite, and which helps to lay hold of and tear off the green food on which they live. The sound of this tearing up the grass is very noticeable when a herd of cattle or sheep

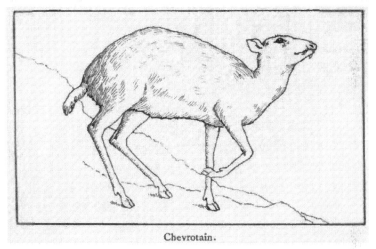

Chevrotain.

are grazing together. But the Camels and Llamas, though they are also Ruminants, have two upper front teeth.

CAMELS.—Camels have been kept for use by dwellers in the East from the earliest times, and formed part of the wealth of Abraham. Their broad cushioned feet, well adapted for travelling over the desert sands, great endurance, power of subsisting on small quantities of the dryest shrubs, and of storing several days' supply of water in their stomachs, render them invaluable to Eastern travellers. They stand about six or seven feet high at the shoulder, and are of a light brown colour, closely resembling that of the desert sand itself. To be loaded or unloaded the camels kneel down, and this is also their attitude of rest, their weight being supported in that position upon thick pads with which the knees and breast are furnished.

Their long soft lips, the upper one of which is split up in the centre, and their slit-like nostrils, give them a curious expression of countenance; but the most remarkable point about the camel is the hump, which acts as a reserve store of nourishment, being large and plump when the animal is well fed, and gradually absorbed during a long journey with scanty food. The Arabian Camel, which is found throughout North Africa, and as far east as India, and the Dromedary, a lighter and swifter variety of the same, have only one hump; but the Bactrian Camel, a native of more northern districts of Asia, has two. The humps are soft and flexible, and blow to one side in a wind. In our recent North African campaigns a Camel corps was raised for the work in the desert; no new feature in war, for we hear that, in David's Amalekite raid, "there escaped not a man of them, save four hundred young men, which rode upon camels, and fled" (1 Sam. xxx. 17).

All the Camels bear a bad character with those who know them well. They are described by a traveller as "great, grumbling, groaning, brown brutes;" and he adds, "never do I remember to have seen a camel in a good humour."

LLAMAS.—In the New World the Camels are replaced by the Llamas of South America, animals of the same family, but much smaller, the largest not exceeding three feet six inches at the shoulder, without humps, without the broad cushions of the feet, and with more sheep-like faces. They have a very unpleasant habit, when annoyed, of discharging the contents of their mouths over the offender, and have been known in this way to rid themselves of their riders when tired of carrying; but they are now rarely used as beasts of burden. The wool of

Llama.

these creatures is long and silky, and under the names of Llama, Alpaca, and Vicuna, according to the species, is well known in manufactures.

On looking back over the numerous Families of this great Order it is interesting to see that in the Bible history notice had been taken of their main divisions, and the only animals reputed clean for food for the Israelites were the cloven-hoofed ruminants. "Whatsoever parteth the hoof, and is cloven-footed, and cheweth the cud, among the beasts, that shall ye eat. The camel, because he cheweth the cud, but divideth not the hoof; he is unclean unto you. And the swine, though he divide the hoof, and be cloven-footed, yet he cheweth not the cud; he is unclean unto you " (Lev. xi. 3, 4, 7).

We do, in fact, reckon the camel among the cloven-footed, but the division of the foot is less marked than with some of the Order.

Rodents (*Rodentia*).—We come now to an Order of animals small in size, but immensely prolific in numbers, and very distinctly marked off from all others, which are aptly known as Rodents, or Gnawing creatures. They have two long front teeth above and two below, with a considerable gap between these and the back teeth or grinders, and no canines at all. But their peculiarity is that the long front teeth are always growing and are only kept to a reasonable size by constant use. Gnaw they must, or they will die; and instances abound in which by the accidental loss of a tooth, the opposite tooth which should have been ground down by it has grown on right through the other jaw till the poor creature could no longer open and shut its mouth, and has died of starvation.

The Rodents consist of four main groups, the Squirrels, the Rats and Mice, the Porcupines, and the Hares and Rabbits.

SQUIRRELS.—Our pretty little English squirrel is a good example of its Family. With its large bushy tail cocked up over its back, its bright eyes, quick, lively, and playful movements, scampering up and down trees, taking flying leaps from one to another, challenging its companions to race, or sitting up nibbling a nut held between its fore paws, it is one of the most captivating of our wild animals. It is eight or ten inches in length, the tail adding seven or eight more inches; it has hind legs much longer than the fore legs, and is in summer of a red-brown colour above and white below, becoming greyer in winter. It is

Flying Squirrel.

a prudent little creature, and, when food is abundant, lays by stores for winter use, for though it sleeps away much of the winter, yet it rouses from time to time to satisfy its hunger. Squirrels of different species are distributed over almost all the world, and a large group of them, the Flying Squirrels, have a deep fold of skin stretching along each side of the body, connecting the fore and hind legs, which is widely extended when they are jumping, and serves to support them in the air like that of the Colugo, or Flying Lemur. One of the Flying Squirrels, the wee little Assapan of North America, does not measure more than four and three-quarter inches in length without the tail. In America also are found most of the Ground Squirrels, which, instead of building their nests in trees, burrow into the ground; and these lead by an easy connection to their relations the Marmots, which are not unlike squirrels with very poor tails. The Alpine Marmot of Europe is large for a Rodent, measuring twenty inches to the root of the tail, but the most interesting of the set is the Prairie "Dog" of North America, so called from its little quick cry like the barking of a small dog. The Prairie "Dogs" live together in great numbers and their burrows are as thickly congregated as those in any rabbit warren, while in front of the mouth of each is thrown up a hillock of the excavated earth, on which the occupant habitually sits. It is a quaint sight to see, from the railway trains running through the prairies, these hillocks, clustered by hundreds, each with a little animal seated on the top of it. Another very curious feature of a "Dog town," as it is called, is that the burrows are constantly shared with a kind of small Owl, known as the Burrowing Owl, about the last creature one would expect to find underground, and sometimes a less welcome guest appears in the Rattlesnake, which feeds upon the young "dogs."

To the Squirrel Group also is referred that intelligent carpenter and builder, the Beaver, which haunts the rivers of North America, having now almost entirely disappeared from Europe, where it used to be plentiful.

Beavers are about two and a half feet long, with a flattened trowel-like tail, and webbed hind-feet. They live together in communities, building their strong lodges in the water, and for this purpose they actually cut down trees by gnawing round and round the trunk with their powerful teeth. The trees, some of which have been measured not less than eighteen inches

Beaver.

in diameter, are neatly cut into logs about five feet long, and then built one upon another with a plaster of mud. The dwellings, when complete, stand out above the top of the water, but the entrances are always under water, and if the beavers cannot find a sufficiently deep pool for their needs they build in the same way a dam across the stream, to pen back a good height of water. Some of these engineering works are of astonishing size, beaver dams having been seen three hundred yards long, and ten or twelve feet wide at the bottom, narrowing up to the top of the water.

RATS AND MICE.—Whatever other animals may be strange to us, we are all familiar enough with rats and mice, and probably regard them with no friendly feelings, from their habits of making themselves at home without invitation, and making free with all eatable property alike in our houses, ships, barns, ricks, and in the open fields. They are found in all parts of the world and in many species, but our domestic rats and mice are characteristic types of the whole group.

The common Brown Rat is a masterful creature, and allows no other species to remain where he has taken possession, so that he has partially exterminated the Black Rat which formerly abounded in this country. Although every one's hand is against the rats, yet they hold their own, partly through the extraordinary rate at which they multiply, breeding several times a year, with from ten to fourteen young in each litter, partly through their power of acting together.

Rat.

They will eat anything, and their disposition varies with their food, but those that frequently get animal food (like the sewer rats in our towns) are fierce creatures, and, when hungry, positively dangerous. A large party of famishing rats would soon pull down a man, and even a single rat will sometimes attack a young child. However, even these unpleasant animals have their use in devouring offal and other pestilence-breeding matters thrown into the sewers; and the country rats are quite content with stealing vegetable food.

The audacious little mice imitate their bigger cousins in the matter of thieving; and they are also very wary creatures, soon understanding and avoiding a trap which has been several times set for them. Mice are undoubtedly fond of music, and individuals among them have some power of singing and even of imitating the song of a special bird. They are quite as prolific as rats.

The Dormouse and the tiny Harvest Mouse, only five inches in total length, are attractive little creatures, and the Harvest Mouse makes a beautiful cradle for its little ones of grass blades woven together into a hollow ball about the size of a cricket-ball, which is slung between the stems of the growing grass by using some of the blades as they grow, without detaching them.

Jerboa.

One of the most remarkable of the rat group is the active little Jerboa or Jumping Mouse of the African and Arabian deserts.

It is six inches long, with about eight inches more in the tail, and its special distinction is the disproportionate length of the hind legs. As it walks and jumps only on these, carrying the short fore paws pressed close to its breast, it has very much the air of a bird hopping.

PORCUPINES.—The peculiar feature of the Porcupine is the number of stiff bristles, spines, and quills mixed with its hair : and when it is irritated, and sets these up on end, it presents a formidable appearance. Its mode

Porcupine.

of attack is by backing upon its enemy with the points of its quills, and as the quills are loosely attached to the skin, and readily come out when touched, they remain in the flesh of the opponent and make severe wounds. A tiger has been found with porcupine quills sticking in its paws and head. The common porcupine is from thirty to thirty-six inches in total length, and is an inhabitant of South Europe and North Africa. Other like species are found in Asia, while the American porcupines are different in their habits and live in trees.

To the Porcupine group belong a number of small

rodents—Chinchillas, rather like squirrels with very soft, grey fur; Agoutis, small, slender-limbed pig-like creatures, very quick and active in running and springing, which do great damage in the sugar plantations of South America; and Cavies, of which the best known in this country is the dull little tailless guinea-pig. In their South American home, however, much larger kinds of Cavy are to be found, and, indeed, the Capybara, or Water Pig, the largest of the Rodents, measuring four feet in length, belongs to this Family.

HARES AND RABBITS.—The Hares and the Rabbits form a single family belonging to the fourth group, and there is a great general resemblance between them, but the hind-legs of hares are nearly twice as long as the fore-legs, while there is much less difference between those of rabbits : the hares' ears, also, are much longer in proportion. Hares live *on* the ground, concealing themselves among the grass : rabbits burrow underground. Young hares are born with their eyes open, and clothed with hair, while rabbits are born blind and naked. They all feed principally in the twilight, but, while the hare loves to lie quiet in the daytime, the little rabbits are constantly engaged in comical play together near their burrows. They are native in almost every part of the world except Australia, and rabbits have been introduced there in recent years; they have multiplied exceedingly and become a great nuisance to the colonists.

Edentata.—The next Order is a complete contrast to the last; for, whereas the rodents are characterized by their large strong front teeth, these have no front teeth or canines, and some of them have no teeth at all, from which they get their name of Edentata, or toothless. They belong entirely to tropical countries, and include some of the oddest beasts living on the earth.

SLOTHS.—The first group of them are the Sloths, hairy animals, without tails, about two feet long, with three toes on the hind-feet, and either two or three on the

Sloth.

fore feet, all armed with long sharp curved claws. Their special peculiarity is that they spend their lives clinging to the under-side of tree boughs ; they travel in this way from tree to tree upside down, they eat upside down, they sleep upside down. But to counterbalance the awkwardness of this position they have an extraordinary power of turning their heads right round, so as to look at anything behind, or beneath them. If placed upon the ground they drag themselves slowly and awkwardly, not upon their front paws, but upon their elbows, to the nearest tree, or anything they can hang to. In the trees they move faster, but are never very nimble, and spend much of their time in sleep, feeding chiefly at night on leaves and twigs. They belong to Central and South America.

ANT-EATERS.—Next come the Ant-eaters, which are strangely various in shape and appearance. They all

Great Ant-bear.

have very long, round, worm-like tongues, which can be thrust far out of the mouth, and which wriggle as if they had an independent life, but this almost seems the end of their likeness. The Tamandua and the two-toed ant-eater of South America are not unlike sloths with long, useful tails, which they twist round branches to hold on by like the American monkeys. Then we have the Great Ant-bear of South America, the oddest of all. It has a very long narrow head, a still longer tongue, a long neck, and is in all about four and a half or five feet long to the root of the tail; but then comes at least three feet more of tail, and such a tail—huge, bushy, plumy—making a complete shelter from the sun when the owner turns it over his back, and lies down under his own shadow. He is a slow, stupid animal; but when very hard pressed will sit up and endeavour, like a bear, to squeeze his enemy to death in his grasp.

Except the Tamandua and its relations, which hunt for insects under the bark of the trees, all these creatures live by scratching down the sides of great ant-hills with their powerful claws, and laying their long sticky tongues among the ants, which adhere to them, and are thus drawn into the mouth of the ant-eaters.

The CAPE ANT-EATER from Africa, which belongs to quite a distinct family, more nearly resembles a pig, if we can imagine a pig five feet long, including its twenty

Pangolin.

inches of tail, without teeth, with a tongue like a worm, and with long pointed ears like a hare. A third set, the PANGOLINS, or Scaly Ant-eaters, form another distinct family. They are covered all over with a complete armour of large, horny, pointed scales, overlapping each other like tiles, with their pointed ends towards the tail. These extraordinary creatures vary from two to five feet in length, most of which is in the tail, and when attacked, roll themselves up so as to protect their heads under their armour, and set up the sharp edges of their scales towards the enemy. They are found both in Africa and in Southern Asia, and are absolutely without teeth.

ARMADILLOS.—The Armadillos, the last Family of the Edentata, all belong to South America, and are dis-

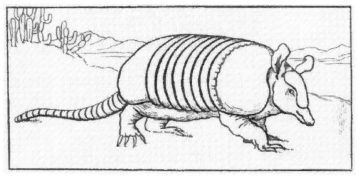

Armadillo.

tinguished by the hard bony shields, like those of Crocodiles, which cover the upper side of their bodies and heads. They vary in size from fourteen inches to over three feet, and have round bodies, short legs, and large, strong claws, with which they burrow rapidly into the earth. They are quick runners, will eat anything, and are themselves very good to eat. One small species, the Ball Armadillo, has the power of rolling itself up, and thus baffling the monkeys, who love to drag back an Armadillo by the tail as it runs to its burrow, but who can make nothing of a ball with nothing to pull at, and too large to be cracked. It makes a beautiful ball, the neat fit of its shields being only rivalled by the beauty of their ornamentation.

Marsupials (*Marsupialia*).—Pouched animals. The next Order of the Mammals, which belongs almost exclusively to Australia and the neighbouring islands, is a very interesting study, for it appears to include animals closely resembling members of many families already described, Cats, Dogs, Bears, Squirrels, Rats, Hares, etc. ; but the Australian forms of these creatures are distinguished by a peculiarity so marked that all who possess it must be referred to one Order. This peculiarity is the history of the birth and nourishment of the young.

KANGAROO.—The Kangaroo is the typical animal, the description of which will best serve to introduce this leading feature of the whole Order.

The male of the Great Kangaroo is a very large animal, clothed in thick warm fur of a greyish brown colour, with a gentle-looking face, large full eye, and upright ears. Its limbs appear much out of proportion, for the fore-legs are short, while the hind limbs are very long, large, and strong, the great hind feet sometimes nearly as long as the leg bone, and armed with claws, one of which is larger than the rest, and a truly formidable weapon. When moving slowly or feeding, the Kangaroo goes on all fours, with an awkward gait ; but when speed is required the hind legs alone are used, and the animal progresses by great leaps, clearing often fifteen feet or more at every bound, and distancing even good horses. It habitually sits or stands upright, supporting itself on the hind legs and great tail, and in this attitude often reaches the full height of a man.

Though the front limbs are short they are capable of much more varied movement than those of many creatures, and may almost be called arms. They can be turned freely on the elbow like a man's arm, can be

Kangaroo.

moved up and down, or put behind the back, and the paw can grasp, hold, or pick up things like a hand.

The Kangaroo will run away from a hunter if it can, but if hard pressed will fight to the last, and is not to be trifled with, as one blow of its powerful hind claws can rip open the body of a dog. When turned to bay by dogs near water it has been known to seize a dog in its arms, hop off to the water, and hold it under water till it is drowned.

91

The female, who is much smaller, carries the distinctive mark of the Order, a large outside pouch of skin on the lower part of the body. When the young Kangaroo is born, it is very tiny and in very undeveloped condition, hardly more than an inch long, colourless, and almost transparent. The mother immediately places it in her pouch, containing the teats, to one of which it attaches itself; and in this living cradle it dwells and grows for eight months, until it is able to take care of itself. As it grows larger, it may be seen peeping out of the pouch; and even when able to come out and feed on the grass, it still runs back to its shelter for rest and safety, until it reaches a weight of about ten pounds, when the mother finally turns it out.

There are many species of Kangaroo, varying in size and in details, and to the same Family belong the Kangaroo Hare, and the Kangaroo Rats, or Potoroos, which are about as large as rabbits, and have heads and teeth somewhat like rodents.

Among the many other curious creatures of the Order, must be mentioned the Wombat, a burrowing animal, with round fat body two or three feet long, a stumpy tail, short equal limbs, heavy waddling gait, and a singularly placid and apathetic disposition; the Koala, or Australian " bear," a little bear-like tailless creature with thick fur, which early transfers its young one from the pouch to its back, and habitually carries it there; the Cuscus, somewhat resembling a Lemur, with long prehensile tail; and the Phalangers, some of whom have a close likeness to the Flying Squirrels. The cats find their Marsupial representative in the Ursine Dasyure or

Native Devil of Tasmania, a savage little creature which commits great havoc among poultry and small animals ; and the dogs in the Thylacinus or Zebra Wolf, also a Tasmanian animal which attacks and hunts down sheep. Both these creatures have great canine teeth, and other points of likeness to the Carnivora.

The only Marsupial animals found outside the Australian

Opossum, with young on her back.

district are the stout furry Opossums of America, the largest of which is equal to a large cat in size, and the smallest not exceeding five inches without the tail. Their tails are a great feature—long, round, flexible— now holding on to the branches to steady their owners in their bird's-nesting excursions, now supporting their whole weight as they hang from a bough, and again arched over a mother's back with the tails of a whole young family twisted round it to secure their safe seat as they ride. The Opossum will eat anything, and is much hunted, both on account of its depredations among poultry, and for the value of its fur.

Egg-laying Mammals (*Monotremata*).—The list of the Mammals closes with an order which may really be considered a sort of link between other mammals and reptiles. The best known member of this very remarkable group is a flattish animal, reaching at most eighteen

Duck-billed Platypus (*Ornithorhynchus*).

inches in length, with a broad, flat tail, like a beaver, flat, webbed feet, furnished with claws, the web of which is folded back when the claws are used for digging; and in place of a snout a broad bill, like that of a duck, which, duck-like, it thrusts into the mud, searching for food, and from which it takes its name of Duck-billed Platypus. Its burrows are made in the banks of streams and pools, the entrances to them being always under water; and much of its life is passed in the water, but it can also run on land and even climb. A gentleman who has kept them in captivity says that they are very cleanly in their habits, tending their fur as carefully as cats, and that the young are very lively and playful.

The Spiny ant-eaters (*Echidna*) agree with the Platypus in their general structure, and in the young being hatched from eggs, not born alive. They are, however, very different outwardly, being provided with a long pointed " beak," and covered with short, strong spines. Both the Platypus and Echidnas are found in Australia and Tasmania, and Echidnas in Papua as well.

All the known Mammalia are included by naturalists under one or other of the fourteen Orders described above ; but in classifying and arranging animals, attention is always chiefly directed to the examination of the bony skeleton, rather than to the outward appearance, so that without a good deal of study it is not always easy to see the reasons why certain animals are reckoned in one family rather than another. It is noticeable that among the lower species of the Edentata, the Marsupials, and the Monotremes which are purposely set last in the series, some details of the skeleton show an approach to characters which belong more generally to birds or to reptiles, and so fitly lead to the consideration of these other great classes of the Vertebrate Animals.

CHAPTER IV.

BIRDS (*Aves*).

BIRDS, which form the second class of the Vertebrate Animals, have several distinguishing characters, marking them off very clearly from all other creatures. They do indeed share with Mammals their red, warm blood, being, in fact, hotter than the Mammals themselves, but from these they are entirely separated by their hatching out of eggs ; and their clothing of feathers and possession of horny beaks are easily recognized outward characters which belong to Birds alone. But the great privilege of birds is the power of flight, for the sake of which their forelimbs are modified into wings. Their bodies contain air-chambers, and even the bones in many birds have cavities filled with air. It is true that there are birds, like the ostrich, which cannot fly, perhaps having lost the power through long disuse, but still the wings exist in a small and stunted form, showing that the same ground plan runs through the whole Class.

Birds annually shed and renew their feathers, but they lose them gradually, so that though they pass through a time of shabby plumage at the moulting season they never become actually bare. They vary much in their powers of voice ; some birds uttering only harsh screams, or twittering or chattering sounds, while others, chiefly belonging to one Order, fill the air with their musical songs.

Another most notable thing in many birds is the extraordinary instinct of migration every year from one climate to another. It is indeed only made possible by their power of flying over the sea, but it is wonderful how unerringly they guide themselves from land to land, many of our bird friends, after wintering in Africa, returning even to the same nest which they built here the previous summer. Among the most noticeable of our summer visitors are the cuckoo, the swallow, and the nightingale, but the migratory instinct is found in birds belonging to almost every Order. The whole subject of the migrations of birds is far from being well understood, and many more observations are needed. There is indeed much of all kinds still to be learnt even about our most ordinary birds, and any one who will observe closely and record accurately what comes in his way, can be of service in adding to our knowledge.

The young of birds are hatched out of eggs, the part of the egg which develops into the young bird being a tiny germ speck. If an egg is laid on its side, the germ spot is always in the middle of the upper side, and may be seen by breaking open the shell at that point. It can be preserved alive for a short time in the same condition, but will only develop into a living bird under the influence of continued heat. This heat is generally supplied by the mother bird sitting on the eggs; but some kinds are hatched by the heat of the sun, or by being buried in heaps of decaying vegetable matter—hot-beds, in fact.

SECTION OF EGG AT THREE STAGES.

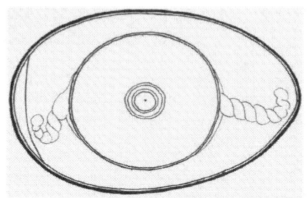

Section of hen's egg when laid.

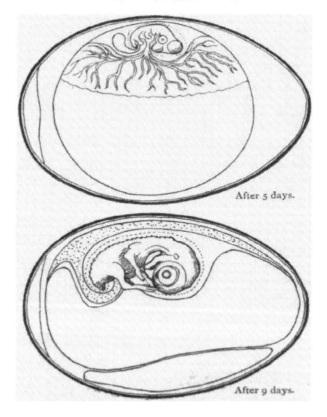

After 5 days.

After 9 days.

The rapid growth in the egg is marvellous to think of. By the end of the first day's sitting, the germ has become elongated and grooved, the brain and spinal cord of the young bird forming along the groove; on the second day the heart appears; by the third bloodvessels have been formed, and so the little chick grows on, supported by the yolk of the egg which is gradually absorbed, until within three weeks the tiny germ spot is changed into a complete bird, which pecks a hole in the shell and comes out into the world. The condition, however, of the young birds when hatched varies greatly in different kinds. Some, like the young thrushes, are naked, helpless little objects, which must be fed by the parents for a long time while their feathers are growing ; while others are like the newly hatched chick of the poultry yard, a dainty little creature clothed in soft down, which will catch a fly for itself with the egg-shell still on its tail.

England is very specially a land of birds. In no other European country, perhaps, are they so continually in sight, forming a constant feature of every country walk, and not infrequent in towns also. Travellers on the Continent must have noticed the comparative scarcity of birds, which have to be searched for, instead of presenting themselves, as here, familiarly at every turn.

Birds of Prey.—The first Order is that of the Birds of Prey. They all have strong, hooked bills, and most of them sharp powerful claws or talons, and they feed on other birds and animals, either hunting and killing for themselves, or acting as scavengers and clearing away what they find dead or dying.

The Falcons and the Owls are typical birds of the two

main groups. The Falcon group includes Vultures, Eagles, Condors, Hawks, and Kites; the Osprey, or Fishing Eagle, holding an intermediate place between them

Condor.

and the Owl group. Owls have soft, fluffy feathers, which enable them to fly very silently, large heads, little or no necks, round flat faces, with eyes looking straight forwards, instead of being set on the sides of the head, like those of other birds, and a general air of composed gravity. Falcons always have three toes turning forward and one

Owl.

at the back of the leg; but Owls can turn the outermost of their front toes either back or forward as they please. The poor Owl is often unmercifully persecuted by gamekeepers in the belief that it destroys young pheasants; but, in truth, it is often accused of the crimes committed by rats, while it really does excellent service by destroying large numbers of rats and mice. Indeed, as the Owls hunt only at night, it is difficult to see how they could get at the young birds, which are then safely sheltered.

Like all birds of prey, Owls throw up in the shape of pellets the indigestible parts of the food they swallow, and an examination of 706 pellets found about a Barn Owl's nest proved them to contain the remains of 16 bats, 3 rats, 237 mice, 693 voles, 1590 shrews, and 22 birds.

Picarian Birds.—The second Order takes its name of Picarian birds from the Latin name of Woodpecker

(*Picus*), which is considered the leading type. They are generally bad nest-builders, and many of them breed in holes. In this Order there are many groups. Some are Climbers which have two toes turned forward and two back, an arrangement with which we are familiar in the Parrots. These beautiful but noisy creatures, of which there are many kinds, are found only in the tropics and in

Woodpecker.

Australia and New Zealand. The last-named country is favoured with several remarkable and troublesome parrots of its own. Parrots have strong, large, curved bills, with which they help themselves in climbing about, and the upper side is jointed so that it can be lifted right up instead of being fixed to the bones of the head like that of most birds. Their mouths and tongues are singularly

102

dry—indeed, a parrot is a dry, powdery bird altogether —and they are continually using their voices screaming, chattering, mimicking, with observant accuracy, and a good memory which has been turned to account in teaching them to talk. It is very difficult to believe that a parrot is entirely unaware of the meaning of its remarks, they are often so quaintly appropriate and humourous. Every one has heard good stories of their conversation ; but few are better than that of the parrot who, having once or twice interrupted the reader during family prayers, was sent out of the room, when, as he was carried to the door, he turned, and said humbly, "Sorry I spoke."

Among Climbing Birds also, we may reckon the foreign Families of Honey Guides, Plantain Eaters, Toucans, and Barbets, and they are represented in England by the Woodpeckers, who climb with their claws and stiff tails while tapping the wood in search of insects, and the Cuckoos who visit us in spring and summer, and who make no nest for themselves, leaving their eggs about in other birds' nests to be hatched.

The Picarian Birds also include Kingfishers, of which there are many more foreign than English ; Hornbills, Trogons, with gorgeous plumage and long tails, Goat-suckers, and others, as well as the dainty little Humming-birds, which hover over the flowers, shining in exquisite colours, the smallest of them having a body hardly larger than a Humble Bee.

Perching Birds is the name given to the third Order, a very large one, including all the songsters, and, in fact, almost all our small birds. Their feet have three toes forward and one behind, all well developed and with claws. First among them stands the Crow group, including Crows, Rooks, Jays, Magpies, etc. Handsome cousins of the Crows are the Birds of Paradise from New Guinea and the neighbouring islands. The body, wings, and tail of one of these beautiful creatures are of a rich brown, the top of the head and neck pale gold, the

Bird of Paradise.

throat and under side of head emerald green, while from under each wing springs a tuft of feathery golden plumes two feet long, which falling backward mingle with the long wire-like feathers of the tail.

Among the Thrushes and Warblers are found our best song-birds ; the Thrush, Blackbird, and Nightingale. The Nightingale arrives in England in the middle of April, and for some weeks until the young are hatched, pours out its song almost all day and night, except for an hour or two in the evening. It does not show itself very freely, but otherwise is not a shy bird, often singing its best by the roadsides. Nightingales seem to answer each other, and may often be started in song by whistling or singing to them. The little Wrens and Tits are members of the same Order.

Another group is formed by the lively Families of the Finches, Swallows, Wagtails, etc. From the time that the welcome Swallows arrive in spring till they assemble together before winter for their departure to warmer countries, they are constantly in sight, sweeping through the air or over the surface of water in search of insects, with rapid and graceful flight, or building round our houses or under our eaves the little plastered nests to which the same birds faithfully return year after year.

Then come the Starlings and their relations, among whom are the cheery Larks; and the last group of Perching birds is entirely foreign, including Bell-birds and Ant-thrushes from America, and the beautiful Australian Lyre bird, which carries its lyre-shaped tail feathers erect like a Peacock.

Game Birds and Pigeons.—To the Order of the Game birds belong those which are most valued for the table, all the varied inhabitants of our poultry-yards, as well as the Peacocks, Pheasants, Partridges, and Grouse. They all have small heads for the size of their bodies, and scaly markings on their feet, and they do not pair with a single mate, but herd many together. The young birds are active and independent as soon as they are hatched, and on this account the Pigeons, whose young are naked and helpless, are separated from this Order in which they were formerly reckoned.

Waders.—Waders come next, and, as their name shows, are mostly to be found near water; but a few land birds are included with them which have the same character of long bare legs, small head, and generally long narrow beak. They are shy birds, migrating

Flamingoes.

annually, and they mostly breed in cold climates, making their nests on the ground.

Rails and Crakes, Moorhens and Plovers, belong to this group, and—though some naturalists place them in a separate Order—we will mention with them the Herons, the Storks—so familiar and highly respected on the European Continent, where they constantly make their nests on the roofs of houses—and the Flamingoes.

No one who has seen them can ever forget the look

of these last, standing, generally on one leg, in the shallows of Lake Menzaleh in Egypt. Troops and troops of them, some five or six feet in height, looking like regiments of soldiers in their scarlet and white plumage, and occasionally bending down their long necks to rake the mud in search of food. They, however, have webbed feet, like ducks, and must therefore be considered as intermediate between the Waders and the Water Birds.

Water Birds.—The Water Birds, who all have webbed feet, are, by some writers, reckoned all together; by others divided into half a dozen different Orders.

Swans are found mostly on fresh water, ducks and geese on both fresh and salt; but the gulls and their

Albatross.

allies belong to the sea, and, from their numbers and variety, form a great feature of life on every sea coast in the world. The grandest of the whole number is the Albatross, whose magnificent wings, when outstretched, cover a space of some twelve feet from point to point. In the southern seas they will follow a ship for many days together, to pick up anything eatable among the refuse thrown overboard, when their unrivalled powers of flight, as they sail sometimes for an hour without a flap of the wings, are a source of the greatest interest to the voyagers.

In strong contrast to these splendid flyers are the Penguins, queer birds on the islands of the southern ocean, whose little wings, quite useless for flight, are modified into flappers, something like those of a Seal. When very hard pressed, they actually use these for running on all fours, like a quadruped; but their usual attitude is that of sitting upright on their tails, if, indeed, they can be said to have tails, and each bird even hatches her

Penguin.

single egg in this position by keeping it close between her legs. The largest species is three feet in height, and they crowd together in communities of many thousands, feeding in the sea.

Wingless Birds.—The last Order of birds is also without the power of flight, but the immense speed at which they can run makes up for this defect. The Ostrich of the African plains is the largest of all living birds, standing from six to eight feet high, with a long neck and long legs, and a considerable likeness, from a little distance, to its neighbour, the Camel. It has two front toes, and, like all the rest of the Order, no hind toe at all. Contrary to popular belief, Ostriches are very

Ostrich.

careful of their eggs, the male birds taking a full share of the work of hatching ; but they have a curious habit of laying additional eggs round the outside of the nest —a mere hollow in the sand—which serve for food for the young birds when first hatched. There seems to be no doubt that, at all events for a short time, an ostrich can keep up a speed of fifty miles an hour, and if pressed in a long chase, will double many times to throw off the hunters.

The birds are valued for their beautiful plumes, and ostrich-farming has now become a regular industry of South Africa.

Birds' Nests.—We cannot leave our feathered friends without a word of admiration for the nests which they

Tailor-bird nest.

prepare with such wonderful skill before the time of egg-laying comes on. At the Natural History Museum in South Kensington you will find a large and beautiful collection of the nests of different birds, forming a

110

delightful and instructive study. The nest-building instinct is doubtless a habit birds have acquired for the protection of their young, and for keeping up the warmth necessary for the hatching of their eggs. And what a wonderful variety of nests are made by different classes of birds! The Ostrich merely scrapes a hole in the sand; Sand-martins, King-fishers, etc., make a burrowed hole; Swallows, Thrushes, etc., make a well-built nest of mud or clay; Eagles and Storks make a flat nest of twigs on elevated spots; most of our singing birds and the crows weave a nest of grass or hair and twigs; the Bull-finches and Humming birds make a soft felt-work nest of wool; whilst the Wren, Titmouse, and Water Wagtails build a covered nest, with an entrance on one side; a few robbers, like the Cuckoo, and the Sparrow if he gets a chance, use the nests of other birds; on the other hand, some, like the Indian Tailor-bird, display almost human intelligence in the construction of their nests. On the last page is a picture of the nest of the so-called Tailor-bird, which is common in the hedgerows of parts of India and China; the name of the bird is derived from the way it prepares its nest. Two or three leaves are actually stitched together by any thread or fibre the bird can get, the bird using its bill to bore the holes for the thread. A sort of cradle is thus formed which the bird lines with cotton wool and fine grass before laying its eggs in its tailor-built nest.

CHAPTER V.

WE now pass to the Third Class of the Vertebrate Animals—the Reptiles, or Creeping Things. Their blood is cold, their breathing and digestion slow, their eyes cold and without expression, and most of them are rather sluggish in movement. They show wonderful tenacity of life, many of them being able to endure fasts for months and even years, and they are often very hard to kill. The young are always produced from eggs, even though, in a few instances, these are hatched before leaving the mother. Whether it is from the unpleasant, clammy feel of these creatures, or from the deadly powers which some of them possess, or from causes less easily assigned, there is no doubt that most people feel an involuntary repulsion from them. They are most numerous in hot countries, diminishing in number as we go north and south from tropical regions; in England, the Class is represented only by a few lizards, three snakes, and some frogs, toads, and newts.

Tortoises.—The first Order is that of the Tortoises—creatures with four limbs and without teeth, in whom the bony skeleton, instead of being wholly covered by flesh, comes partly to the outside in the shape of large, bony shields, which form a protecting armour over the back and breast. Some of the Tortoises are able to treat their shell actually as a house, withdrawing their heads and limbs entirely into them ; and one group, the Box Tortoises, have a jointed piece of shell with which they can close their doors against all enemies, but others cannot thus protect their heads, which are always protruded from the shell. They all lay eggs, digging holes to put them in and covering them over, but they then leave them to hatch out without further attention, and

Tortoise.

do not watch over the young, who are lively and able to take care of themselves as soon as they come out. Many sleep all through the winter, burying or hiding themselves under rubbish, and all become more sluggish when it is cold. There are sea tortoises, including the green Turtles, so much valued for the table, freshwater, mud, and land tortoises, and they vary in size from monsters, five and a half feet long, four feet wide, and three feet thick in the body, to the little tortoise from the south of Europe, which is often imported into

England as a garden pet, and which can lie on a man's hand.

Crocodiles and Alligators are ferocious reptiles, haunting the rivers of hot countries. Crocodiles are found in Asia, Africa, Australia, and America; Alligators, which differ from Crocodiles in their bones and teeth, in America only, except one kind which lives in a river in China. They have flat, long bodies, of a dirty, dark colour, protected on the back with solid scales, and long tails, the whole length sometimes reaching twenty feet

Crocodile.

or more. Their limbs are short and powerful, and their toes somewhat webbed, and they can go either in the water or on land, but are much more nimble in the water. Their long jaws, armed with a formidable array of sharp teeth, can be very widely opened, and as they will devour anything animal that comes in their way, they are often dangerous enemies to men. Considering the size of the full-grown reptiles, they lay strangely small eggs, not larger than those of a goose ; but the tiny crocodiles that emerge from them are very like their parents, and already armed with their sharp teeth.

Lizards.—In the Third Order of Reptiles we find all the many varieties of Lizards, of which our common little English lizard may be considered a good representative. It is about six inches long, with a slender, scaly body and long tail, a long, forked tongue, and fair-sized limbs, with five widely spreading toes. In warm weather it loves to lie basking in the sun, or darts about in a lively manner catching flies. It is one of those whose young are born alive.

All the Lizards are rather apt, when handled, to snap

Lizard.

off their tails, which, however, grow again, a peculiarity which marks them as animals of a low order. They usually have four limbs, but in some cases, as, for instance, in the slow-worm, no limbs are visible outside, only traces of them being found under the skin. This makes them look very like snakes, from which, however, they are distinguished by the shape of the head and manner of opening the jaws.

There are Lizards of all sizes, the Nile Monitor, which is the largest, measuring six feet in length. Some live in water, some on land; some have long forked tongues, some thick fleshy ones just notched at the tip; several change their colour under different circumstances, of which the Chameleon is the most noted example; and some, like the Gecko, have flattened feet, like suckers, which enable them to run on upright and slippery surfaces.

Snakes, the most formidable Order of Reptiles, are not conspicuous in this country, but in some hot climates

Viper.

they constitute a serious danger. The outward form of a Snake is well known, elongated and slender, covered with scales and without limbs, travelling, often with great rapidity, by a sort of gliding movement, extended along the ground, but able to lift up the head and fore-part of the body. But the speciality of Snakes is in the poison fangs possessed by many of them.

These are like two long teeth in the upper jaw, curved and pointed downward when extended to strike. They contain hollow tubes, and when a wound is made by their sharp points, a drop of colourless venom is squeezed through them from the poison bag which lies behind their base. The power of this venom varies in different species; thus, the bite of the Viper, the only venomous Snake

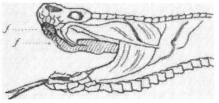

in England, though it kills small animals, is rarely fatal to men, unless they are already in an unhealthy condition; but the bite of the Indian Cobra, the

Head of Venomous Serpent showing Fangs *f f.*

African Puff Adder, and many others, will kill very rapidly. A Viper may always be known by the zigzag chain of dark markings that runs down the spine, while our common harmless ringed Snake is darkest on its under side and of a lighter greenish grey on the back. Snakes swallow their prey whole, the bones of their jaws being loosely jointed together, and so far separable that they, as well as the throat, can be enormously distended, and allow of the passage of objects which might beforehand be thought far too large to go down. All Snakes are not venomous, but some of the Python group, which have no poison fangs, are of very large size, sometimes eighteen or twenty feet long, or even more, and they kill their prey by winding their coils round it and crushing it to death. Even animals as large as deer are swallowed by these serpents.

Amphibia.—This is the place to speak of the Amphibian animals, which have been sometimes included under Reptiles, but are now always placed in a separate Class. Their peculiarity is that, when hatched out of the egg, they are quite unlike their parents, and only by a series of changes gradually acquire the same form. The best known of them are Frogs and Toads, which are born as tadpoles, little dark creatures swimming about in water, with a large flat tail but no limbs, and the heart and breathing apparatus of a fish. By degrees the limbs begin to grow, the hind legs appearing first and then the front ones; and as the limbs increase in size, the tail gradually disappears, not dropping off, but being absorbed into the body.

During these changes the little creature is also developing lungs fit to breathe atmospheric air, and as soon as

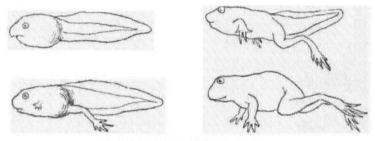

Tadpole and Frog in different Stages.

the tail is gone it comes ashore a perfect little frog or toad, and henceforward spends much of its time on land, though always loving moisture and haunting cool, damp places. They have shiny skins, without scales, and remarkable tongues, which are fastened to the *front* of the lower jaw, and lie with the tip pointing down the throat: and consequently can be protruded to a considerable distance when shot out in pursuit of insects. Frogs are the more active creatures, moving usually on land by hops and leaps, while toads crawl. They can be tamed; and toads especially, which are often kept in gardens and greenhouses to destroy insects, soon come to know those who are kind to them. Toads secrete a kind of acrid juice in their skin, which makes them very distasteful to dogs and other animals, and this has probably given rise to the idea that they are venomous, but they have no real venom.

There are other Amphibian creatures that never lose their tails, as the Salamanders and the common Newts, or Efts of our waters; and a word must be said of an extraordinary Mexican Amphibian, whose eggs hatch out sometimes what we may call the tadpole form, but sometimes the mature form at once, with all its organs complete. Moreover, to complicate the matter, eggs are laid both by the complete creature and by its tadpole form, and there is no telling which form of the creature will come out of the eggs of either. The immature animal rejoices in the name of Axolotl, while the mature is known as Amblystoma.

CHAPTER VI.

FISHES.

THE last Class of Vertebrate creatures is that of the Fishes, which are cold-blooded and live entirely in water, breathing through gills the air contained in the water. A fish out of water dies when the gills become dry, but two or three species, such as eels, and a wonderful fish called the Climbing Perch, have gill covers able to

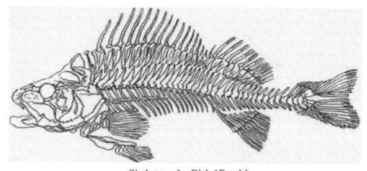

Skeleton of a Fish (Perch).

retain a store of water to moisten the gills; these can maintain life out of water as long as the store lasts, and are thus enabled to pass from one pool to another.

Most fishes are covered with scales overlapping each other, but in some the place of the scales is taken by bony armour. The fins of a fish are its nearest approach to limbs, and the two front pair (pectoral and ventral), where they exist, may be taken to represent the fore and hind limbs, but the back fins (dorsal), the tail fins (caudal) and the anal fin (see illustration), have nothing that corresponds to them in the higher Vertebrate animals.

Fishes are produced from eggs which, like those of reptiles, are sometimes hatched before birth, and their

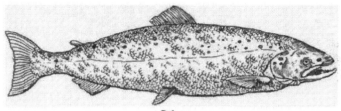

Salmon.

fertility is extraordinary, the number of eggs which form the roe of a single fish being sometimes counted by millions. The eggs and young are generally left to take care of themselves, but a very few fishes, among whom are some of the Stickle-backs, build nests and watch over their young.

If we take the Salmon as a type of the fish form, we

Shark.

may notice how beautifully it is built for rapid motion, like a swift ship, its sloping lines offering the least possible resistance to the water. It is driven forward by strokes of the tail, the fins helping to balance and steady the

121

body. But in truth, though the general plan of fish shape is very distinct, there are really endless modifications among different kinds. Thus there are fish with their mouths at the end, and others with mouths under-

Sunfish.

neath their heads, like the Shark family, the larger members of which are the terror of the sea, while others again have long snouts, like the Pipe-fish, or formidable weapons projecting from their noses, as the Sword-fish. The great Sunfish, which sometimes reaches a length of

six feet, is almost round, and looks rather like the head
and shoulders of a huge fish which has lost the rest of
the body. Skates are flattened out as if heavy weights
had been pressed on the back of a very wide fish, and
had squeezed it into a sort of resemblance to a child's
kite : and on the other hand there are a whole group of
fishes, squeezed in from side to side, which are in the
habit of lying flat on the bottom of the sea, and so have

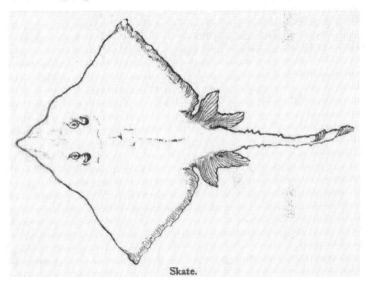

Skate.

both their eyes twisted on to the same side of their head,
that neither of them may be underneath as they lie.
These flat fishes are very excellent food, as they include
the Turbot, Plaice, Flounder, Brill, and Sole. But the
most important fisheries are those of the Cod, which
are found most plentifully off the coasts of Newfound-
land, and of the Mackerel and Herring, which at certain
seasons assemble in vast shoals, and come towards the

shores for the purpose of spawning, or laying their eggs, followed by many enemies in the shape of birds, large fish, and Cetaceans, as well as by the fishing-boats. It

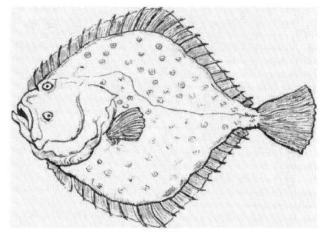

Flatfish.

is a pretty and interesting sight to see these boats come in and unlade their catch, the Mackerel in particular, being a beautiful fish coloured dark green, black, and silver. Sprats and Anchovies are nearly related to the

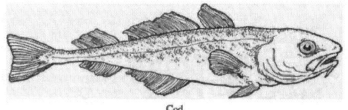

Cod.

Herring, and there are plenty of other useful sea-fish, while the fresh waters supply Trout, Carp, Tench, Barbel, Perch, etc., and there are both fresh and salt-water Eels. As a rule, the fresh and salt-water fishes keep each to

124

their own domain; but the Salmon migrate annually between them, coming up the rivers to lay their eggs, and returning to the sea to recruit their strength after the exhaustion of this process. It is considered that the migration also helps to free them from parasites, of which the river kinds are killed by the sea, and the sea kinds by the river water.

CHAPTER VII.

INVERTEBRATE ANIMALS.

IF it has only been possible to give a very slight sketch of the main Orders of the Vertebrata, what can be said of the vast families of the Invertebrate creatures, of

Octopus.

which a single Order, that of the Beetles, has been estimated to contain more than double the number of species of all the Vertebrate animals put together, and in which the individuals are simply countless myriads, apparently becoming the more numerous the smaller their size?

Mollusca.—Of their main Divisions the first is that of the Mollusks, creatures with soft bodies, either naked, or covered in whole or in part with shelly covering. They have a mantle, or soft membrane surrounding the body, by means of which most of their movements are made : and the higher mollusks have a kind of heart and blood vessels, but these cannot be distinguished in the lower.

The Octopus and Cuttle-fish have distinct heads, round which are set long tentacles covered with suckers by which they grasp their prey. Some of these creatures are of enormous size, well authenticated cases being known in which the tentacles or arms have reached forty feet in length. Large specimens are dangerous enemies, for whatever they grasp they hold on to with tremendous grip, and men seized by them have only escaped by cutting the tentacles to pieces.

Another group of the Mollusks contains snails, slugs, and the shell-fish that form what are called univalve shells, that is, shells made all in one piece like a snail-shell. These are often extremely beautiful in their colouring and marking, and very various in shape, but they almost all have a tendency to be spirally coiled up, more or less tightly. Some, when alive, have their mantles wrapped over part of the outside of their shells, but many can withdraw themselves entirely into their shelter, like the snail, who only puts out his great flat, crawling foot, and his head with its eyes carried on footstalks, when he is satisfied that no danger is near.

Bivalve shells, which form another group, are those made in two pieces and hinged together, like the Oysters, Solens, and Scallops. The greater number of their inhabitants are very sedentary, often fixing themselves to

Whelk.

Scallop.

one spot for their whole life. Pearls are found in the shell of a sort of oyster, and other bivalve shells furnish the beautiful substance mother of pearl.

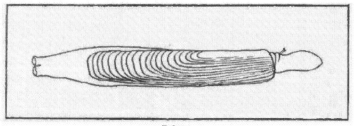

Solen.

Arthropoda.—The vast Division called Arthropoda includes all that have jointed legs among the animals whose bodies are arranged in successive rings or segments. They may be divided into four Classes.

1. The *Insects*, whose bodies are in three parts, head, thorax, and abdomen : they always have six, and only

128

six, legs, attached to the thorax, or middle division, and two *antennæ*, or feelers on their head; and generally one or two pairs of wings also.

2. The *Myriopoda*, or many legged creatures, Centipedes and Millipedes, which have wormlike bodies in

Caterpillar.

Pupa.

Butterfly.

successive rings, varying in number from ten up to one hundred and sixty, and legs on nearly every segment.

3. *Arachnida*, or spider-like animals, with eight legs, and no wings or antennæ.

4. *Crustaceans,* Crabs, Lobsters, etc., with two pairs of antennæ and many pairs of legs. In some the body segments are very distinct; in others, such as the Crabs, some of them are welded together to form strong shields on the back and breast.

Insects have a specially interesting life history in the number and completeness of the changes which many of them undergo, being hatched from the egg in the shape of maggots, grubs, or caterpillars, which can only crawl and eat voraciously (this is called the *larva* stage); then passing into a second condition called the *pupa,* closed up in a case of skin, motionless and apparently dead for some time, until the case at last splits asunder, and the insect comes out perfect, and generally with wings wherewith to fly about in the air.

Those that go through the whole of these changes are the Beetles, whose front pair of wings are not used for

Beetle with spread wings.

flying, but merely form horny cases or sheaths to put away the flying wings in; the Ants, Bees, and Wasps, and all their relations; the Butterflies and Moths, Caddis

Flies and Lacewing Flies, all the innumerable varieties of two-winged Flies, and the Fleas.

Volumes have been written about the Bees, and their wonderful communities, living together round their queen,

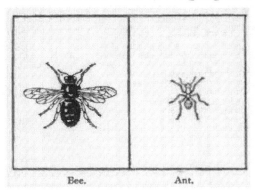

Bee.　　　　Ant.

the mother of the hive, working together for the common good at building their beautiful combs, collecting honey from the flowers to fill them, tending the young, and when the hive becomes too populous, emigrating in a swarm to found a fresh city elsewhere. Not less wonderful are the histories of Wasps, and of the Ants, some of whom actually keep herds of the tiny Aphis, or green fly, often so abundant on rose-trees, and do something very like milking them regularly. You must read about them in larger books.

Butterflies and Moths, the loveliest of Insects, have their wings clothed with tiny feathery scales, which, like works of inlaid gems, form all the beautiful colouring and patterning. The scales differ in shape, and are beautiful objects as seen through a microscope. The Butterflies have long antennæ ending in a little knob, while the antennæ of Moths are pointed at the tip, though often like combs or feathers below. These are the most conspicuous and interesting insects in which to watch the gradual transformations from the caterpillar and chrysalis.

To the Fly group belong all the true Flies, the Midges, Craneflies or Daddy-long-legs, and Gnats and Mosquitoes. These last lay their eggs in water, glueing them together into the form of perfect little boats, which float on the top until the eggs hatch out, when the larvæ fall into the water, where they pass their early days.

Crane Fly.

There remain some insects which do not go through all these changes, or do so only imperfectly, either resembling the parents from the first, or being active in the pupa stage. Such are the Bugs, including Cicads, Greenfly, and Froghoppers ; the Cricket and Grasshopper group, which also includes

Grasshopper.   Earwig, flying.

creatures so different in appearance as Cockroaches and Earwigs, White Ants or Termites, Mayflies, Dragonflies, and Springtails.

The second Class of the Arthropoda, the Centipedes

(see p. 14) and Millipedes, show very distinctly their formation in rings. They do not go through the same series of changes as Insects; but in some Millipedes the young when first hatched have only three pairs of legs, acquiring more at each change of skin, until they fully resemble their parents.

The third Class, which includes the Scorpions and Spiders, have eight legs, and are the only Arthropod creatures without antennæ. They are largely carnivorous, and many of the Spiders make nets to catch their prey, cleverly spinning their thread from a sticky substance, which issues from their own bodies. They vary in size, from tiny creatures that need a magnifying glass to distinguish them, up to monsters whose legs cover a space of six inches, and who are able to attack and kill humming-birds and other small vertebrates.

But the most formidable of the Arachnid animals are the Scorpions, found in most hot countries, whose terrible

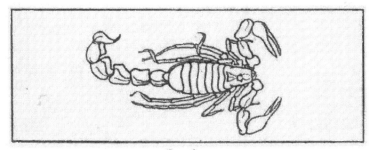

Scorpion.

poisoned sting, though rarely fatal to the life of a man, "puts him," as it has been graphically expressed, "to the necessity of howling for the next four-and-twenty hours."

The Order of Crustaceans completes the number of the Arthropoda. It consists of Crabs and Lobsters, Prawns, Shrimps, "Water-fleas," Barnacles, and such-like creatures, mostly covered with a hard shell or crust. They almost all live wholly or partly in water, except the familiar Woodlice which haunt our gardens, and some Land-crabs. These Land Crustaceans resemble their parents from the time they come out of the egg ; but the water creatures only gradually arrive at the perfect

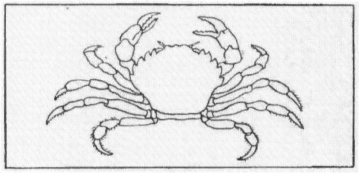

Crab.

form through a series of changes, much more gradual and numerous than those of the insects, and made through successive moultings or castings of the skin.

Worms.—We must pass on to a few words about Worms, an extraordinarily numerous assemblage of animals, of which the common Earthworm (see p. 12) is a good example, so useful in producing good soil, and so delightful to the thrushes and blackbirds that busy themselves in dragging the worms from their holes. They have no limbs at all ; but their bodies are arranged in a series of rings bearing small bristles, by means of which they move with a sort of snake-like motion. Earthworms lay eggs; but they also have the power, if cut in half, of growing a fresh head or tail, so that the one individual becomes two; and in several of the Worm Family this power goes further, and they divide up of their own accord, the tail part growing a fresh head before it separates from the old one, so that for a time the creature goes about with two heads to one tail. One or two species carry this so far as to have half a dozen worms behind them sharing their tail, all of which eventually separate. In this case the old front worm alone does the eating, while the supplementary individuals occupy themselves solely in laying eggs.

Some of the sea worms, though without true jointed limbs, have simple projections that may be called feet; and many form shelly or sandy tubes to live in. A beautiful example of the tube worms is seen in the Serpula, well known to any one who has studied a sea aquarium. From the mouth of its twisted tube it protrudes a plume of breathing gills, which, when fully expanded, forms a brilliant scarlet fan, beside which stands up a footstalk carrying a stopper of the same rich colour. But the Serpula is shy: at the slightest suspicion of danger, sometimes at the mere falling of a shadow across the water, the whole pretty show is gone like a flash of lightning, withdrawn into the shell, the mouth of which is closed with the scarlet stopper.

Besides land worms and water worms, there is a highly disagreeable set of parasitic worms, inhabiting the bodies of living animals, where they often give rise to serious discomfort and disease. Almost all creatures, from man downwards, are liable to attacks from some of the numerous species of Internal Worms.

The three Divisions of the Invertebrate Animals that remain belong entirely to the water, and chiefly to the sea.

Serpula.

Echinodermata.—First come the creatures the parts of whose bodies are set in rays round a central opening, and who are covered with prickles or spines. If any one picks up and handles a Starfish (see p. 14) stranded on the seashore, he will at once see that it belongs to this group. Starfishes are active creatures, walking about by means of the innumerable tiny suckers or tentacles with which its rays are furnished. They are voracious, and do immense damage by devouring the oysters in oyster beds, and often take the bait on the fisherman's hook. The angry fishermen, finding that their catch was only a worthless starfish, used to tear them in half, and throw them back into the sea, a proceeding which made two starfishes in the place of one, for, like most creatures in these lowest groups, each half could heal its wounds and

Sea-Urchin.

grow again what it had lost; even a single arm being sufficient to grow a new starfish.

They lay enormous numbers of eggs, and protect the young until they have developed their rays and tentacles.

To the same Division belong the Sea-Urchins or Sea-Eggs, the upper part of whose bodies is covered with a shell of most elaborate and beautiful construction, armed all over with spines; and also the Sea Lilies and Sea Cucumbers.

Cœlenterata.—The creatures of the next Division

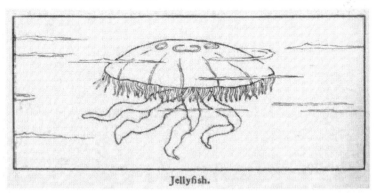

Jellyfish.

are distinguished from all those which have gone before by the absence of a body-cavity between the body-wall

137

and the food-canal. The Jellyfish are like beautiful, almost transparent umbrellas, with tentacles hanging down from the under side, and they float in the water with their umbrellas opening and partly closing in a constant pulsation. They vary in size from tiny microscopic objects to a diameter of several feet, and many of them are not only of extreme beauty by day, but shine with phosphorescent light at night. There are few more beautiful and interesting sights than a night at sea when the phosphorescence is strong; tiny star sparks continually snapping in and out, while among or below them float the pale white moons of the larger Jellyfish. Care should be taken by sea bathers not to come in contact with Jellyfish, for many of them sting like nettles, and a few produce really severe effects.

Sea-anemone.

The pretty Sea-anemones are familiar to all dwellers by the sea side, and a great ornament to the aquarium; bearing a considerable likeness to flowers when all their tentacles are spread out, but when these are withdrawn, subsiding into shapeless lumps of jelly.

Nearly related to the Sea-anemones are the wonderful little creatures that build and live in coral.

Every one has seen coral, dead coral, in white branches like trees, in rounded masses folded in and out like the brain, in clusters like multitudes of wee organ pipes, or in some others of its numerous forms; but few have the chance of watching living coral, when every little tube has its living tenant, and they thrust out through every pore their mouths fringed with tentacles like tiny sea-anemones. Corals grow on the bottom of the sea, or on rocks below the sea level, and increase both by eggs which escape from their mouths, and by budding from the side of the parent. The little creatures—polypes they are called—deposit hard structures from the lime contained in their food or absorbed from the sea water, and in their turn bud again.

The polypes of the older parts die, but the solid structure remains as a foundation below the newer

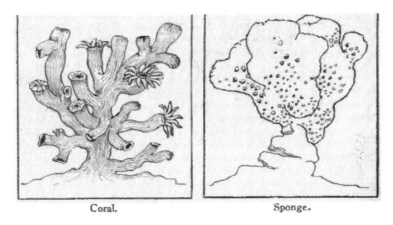

Coral.   Sponge.

growth, and so the whole mass grows until reefs of great extent are formed, generally surrounding islands. The reef-building corals require warm seas, and they are principally found fringing the islands of the Indian, Atlantic, and Pacific Oceans; but other coral polypes, some of which dwell separately and not in colonies, belong to cooler seas, and some are found on our own coasts.

We may conveniently class the Sponges with the Zoophytes and Corals, though they are often reckoned as a group by themselves. The simplest form of sponge is like a hollow bag, with numerous holes through which currents of water pass to the inside, and a mouth at one end through which the water passes out. As the water passes through the sponge, the tiny *cells* which line the inside feed on the little living creatures in the water. In most sponges the inside lining becomes branched into a very complicated system of tubes or canals leading to chambers where the feeding goes on. A hard skeleton of limy, flinty or horny spicules or fibres, is usually developed to support the soft, living tissue of the sponge.

Protozoa.—Lastly we come to creatures which throughout their life are simple cells, as a rule so tiny that they cannot be seen without a micro-scope, but which yet often form as their dwelling places shells of considerable va-riety and beauty ; as, for instance, the mi-nute sea-shells which are known as Fora-minifera. These shells are formed of lime, and though each of them is not bigger than a pin's head, yet their accumulation in

Living Foraminifer (much magnified).

past ages has formed the great chalk deposits which we all know in the cliffs and downs of the south coast of

England. A living Foraminifer collects and absorbs its food, and is able to move about by means of ever-changing and interlacing threads of living matter which stream outwards from its tiny shell.

As an example of a simple animal belonging to the same group as the Foraminifera, but without a shell, we may take the very elementary freshwater creature called Amœba, consisting of a single cell of slimy substance, soft, naked, constantly changing its shape, without mouth or stomach, but which, nevertheless, contrives to absorb other living creatures by putting forth processes which surround them, and gradually digesting them.

We have now passed in rapid review the main groups of the Animal Kingdom, from Man down to these simple creatures we have just described, which remain all their lives long in the condition of single cells. It is of great interest to remember that any of the higher animals begins its life as a single cell—the egg-cell—from which it is built up by a slow or rapid process of growth. So it is believed that in the animal world generally there is progress from the simple to the complex, from the lower to the higher forms of life.

CHAPTER VIII.

AFTER this brief review of animals in their classes, it will be interesting to examine a little more closely how they are made, what are the different parts of the body, how it is nourished and moved, and how the actions of life are performed; but as there is a great general similarity in construction, we will take the human body, the most highly developed and also the most interesting, as the type, and while giving a short account of it, will try to notice at the same time some of the main points in which it differs from other animals, or in which they differ from one another.

The pillars and foundations of the body are to be found in the bony skeleton, which supports all the softer structures, and which is far more permanent than they, being left when the rest of the body decays.

Look at the picture of the human skeleton on p. 28, and notice first the main central pillar, or spine, or vertebral column, reaching from the central part of the body, where the legs join it, up to the skull. It is formed of a series of small similar bones, laid one on the top of the other, and jointed together by a tough strong substance, so that, though quite firm and strong, it is not stiff like a poker, and we are able to bend it a little when we stoop, or lean to one side. The skull, which rests on the top of it, is a strong, solid bony case, hollow inside, with the cavities of the face on its front surface, and the lower jaw, which is a separate bone, jointed on to it.

If the small bones, or vertebræ of the back, are separately examined, it is seen that each consists of a solid central portion from which springs a bony arch enclosing a hole and bearing projections. When the vertebræ are laid over each other, the holes form together a continuous tube passing right up the centre of the column, enclosed in a strong case of bone, and opening above into the large hollow of the skull. The skull and the tube together are like a strong safe, and they contain and protect a most important treasure which we shall speak of presently.

The jointed backbone, as we know, is the distinguishing mark of all the Vertebrate animals. Look at the skeletons of creatures from different Classes of the Vertebrata—on pages 5, 7, 27, 28, and 39—and you will see that they all have a spine and a skull, but both the shape and number of the vertebræ differ very much in different creatures. You will be interested to notice some of these varieties on the dinner-table in the animals we use for food. In the spine of a pig the massive vertebræ are comparatively simple in shape; but the vertebræ of rabbits are complicated with sharp projections; so are those of a fowl, but we cannot so easily make them out, as all the winged birds have their hinder vertebræ firmly united into one solid mass.

In man there are thirty-three vertebræ, and they come

to an end, as you see, just below the attachment of the lower limbs; but in many animals they are continued into long tails (see p. 39), and some large snakes have more than four hundred, moving upon each other by such free joints that the creatures can coil themselves up into knots (see p. 5).

The seven top joints of the spine in man are above the attachment of the upper limbs, and so belong to the neck; and it is a curious thing that throughout all the Mammalia there are always seven vertebræ in the neck, neither more nor less, except in those odd creatures the Sloths, some of which have

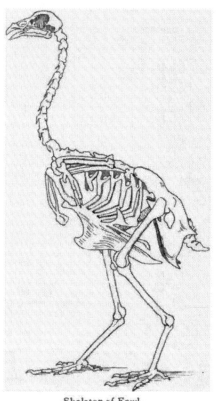

Skeleton of Fowl.

nine and others only six. Even the Giraffe has but seven, and consequently each bone is so long as to make his long neck look very awkward and inflexible.

With birds, on the contrary, the neck vertebræ vary in number from nine up to twenty-three, so that, though their backs are rigid, their necks are very flexible and enable them to reach every part of their bodies and every feather with their beaks.

The twelve vertebræ next below the neck (pages 27 and 28) carry the ribs, long curved bones bending round to the front of the body, the upper part of which they enclose like a cage. The front ends of the ribs do not reach right to the flat breast-bone, and the two lowest pair are very short; but most of them are joined to it by bands of cartilage or gristle, a tough strong substance not so hard as bone. All the Mammals have the breast-bone, or *sternum*, as it is called, more or less flat, like that of a man; but in the flying birds it is large and extended forwards into a sharp ridge or keel, as is well known to any one who has carved the breast of a fowl (see picture on last page). Coming to the Reptiles, we find that some Lizards and Crocodiles have a sternum, flat like those of the Mammals; but other Lizards, Tortoises, and Snakes dispense with it entirely, as do also the Fishes.

The number of ribs varies greatly in different creatures.

Next we come to the limbs, of which there are never more than two pairs in Vertebrate animals, and we find that each pair is attached to and supported upon a sort of bony circle or girdle. In man the shoulder girdle, which extends round and completely outside the upper ribs, consists in front of what is called the collar-bone, or *clavicle*, slightly curved strips of bone joined to each side of the top of the sternum, and meeting on the point of the shoulder with the shoulder-blades, large flattened bones which form the back of the circle.

The upper part of the arm has but one bone, called the *humerus ;* but from the elbow to the wrist we find two, so jointed to the elbow that the arm can either be bent or straightened, and so attached to each other that the arm and hand can be turned in different directions. Lay your hand upon the table with the palm downwards, and you will find that you can turn the palm upwards without moving the upper arm at all, by the rotating upon each other of these two bones, of which that widest at the elbow is called the *ulna*, and the other, which is widest at the wrist, is known as the *radius*. The finger bones are called *digits*, and their separate joints *phalanges*.

The bony girdle which supports the lower limbs and soft lower parts of the trunk is called the *pelvis*. In young children it consists of three bones on each side, but in a full grown person the three are firmly grown together, and the pelvic bones form a flattened irregular sort of basin, to each side of which are attached the leg bones. The bones of the leg correspond to those of the arm, the single strong bone of the thigh bearing the name of *femur*, the two between knee and ankle being called the *tibia* and *fibula*, and the toes sharing with the fingers the names of digits and phalanges.

Now, let us compare the bony girdles and the limbs in other animals. Their general correspondence with those of man is very evident on looking at the skeletons, and we can generally without difficulty name each bone from those of the human skeleton. It is true that Whales and Manatees have no hind limbs, and Snakes no shoulder girdle or fore limbs, yet we find in all these Orders very small pelvic bones, not joined with the spine, but embedded in the flesh, and a few of the Snakes have traces of actual hind limbs.

A principal variation from the human type is in the presence or absence of the collar bones or clavicles. It seems as if these bones, which have nothing corresponding to them in the hinder bony circle, were only necessary in those animals which use their forepaws for something else than just to stand upon. Bats have it well developed to support their wings; but, as the twisting arrangement of the fore arm might make their flight unsteady, they have the ulna reduced to very small proportions, while the fingers are enormously lengthened to carry the wing membranes. Rodents, also, several of whom, like the squirrels, hold up their food in their paws, have good collar bones, and so have Sloths, which swing by their long arms on the tree boughs, and Kangaroos, which use their forepaws almost like arms.

But, in the beasts of prey, the clavicle is either absent or very little developed, and the forelimbs are not directly attached to the bones of the trunk, but are held to it only by cartilages and soft structures, so that a tiger alighting from a leap does not get a great jar through all its frame from the fall, as we should do. The absence of the clavicles is one of the distinctions that marks off the beasts of prey from the Insect-eaters (p. 34).

Elephants, cattle, deer, horses, etc., which do not lift their forelegs high, are not only without collar bones, but the double bones of the limbs are so fixed together, that there is no power of rotating them, while they are strong and steady as supports.

We may be sure that birds, which depend so much on their wings, must have good clavicles ; and, perhaps, we shall recognize them better under the familiar name of the merrythought.

Look back at the skeletons of the gorilla (p. 27), the lion (p. 39), and the snake (p. 5), and compare them with that of the man (p. 28). The gorilla is really not unlike a man, except for its very long arms, receding skull, and prominent jaws.

As to the lion, try for a moment to imagine the skeleton set upright in the same position as the man's. You will notice that its hind legs can never be completely straightened into the same line as the backbone, and its fore limbs can neither be lifted up towards the head, nor dropped down by its sides, but must always stick straight out in front. No other creature but man can really stand and walk upright; some of the apes come nearest to it, but they cannot truly straighten their knees, and always want to pull themselves up by their arms. The lion's skeleton shows well, too, how all the cats walk on their toes, with the heel lifted from the ground, though pussy puts down the whole of her back feet when sitting up in her dignified manner.

Certainly the snake is different with its absence of limbs, for it is just a long series of vertebræ and ribs, with a head and a tail.

Thus we see that the general plan of bones,—skull, spine, ribs, breastbone, and two pairs of limbs with a bony girdle supporting each,—is traceable on the whole, in spite of variations in details, throughout the Vertebrate animals. The widest departure from it is in the

Fishes, the lowest Class of Vertebrata, where there are
additional rows of small bones.

Let us pass on now to look at the machinery by
means of which the bones are made to move in living
bodies. We cannot have a more convenient example
to begin with than the way in which we bend and
straighten our arms. Here is a figure of the bones of

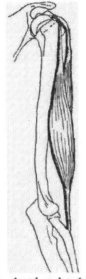

1. Arm hanging down.   2. Elbow bent.

Bones of Arm with biceps muscle.

the arm as we saw them in the skeleton; but now
there is added a certain mass of flesh, wide in the
middle, but narrowing down at each end into firmer
and stronger cords called tendons, by which it is
fastened to the bones. The lower tendon is attached
to the *radius* of the arm just below the elbow, while,
at the upper end, not one, but two tendons pass

upwards, and are secured round the shoulder blade, on to which the arm bone is jointed.

Now this fleshy mass, which is called a muscle, has the power of *contracting in length*, or drawing up its particles so that it becomes shorter and wider, and when it does this it draws up the lower part of the arm, bending it upon the elbow and bringing the hand up towards the shoulder, in the position shown by the outlined arm in the figure. Stretch your arm out quite straight, clasping it round with the other hand about halfway between the elbow and the shoulder, and then, if you bend the elbow, you can feel the muscle rising up under your hand as it grows thicker and shorter.

This particular muscle is called the biceps, or two-headed muscle, from the two tendons in which it ends above, and its work is to bend the arm; but it is by no means the only one in the arm. There is another muscle over the back of our elbows, and when this back muscle contracts it draws the lower part of the arm down, and so straightens it out again. When the arm is held in a half-bent position, each muscle is partly contracted, and the pull of each just balances the other.

Muscle is what we know as " lean " meat or flesh, and, as you know, the bones of almost every part of the body are covered with this soft muscle. It does not, however, lie anyhow, like mere padding, but every bit of it is a separate muscle which has its own definite direction in which it contracts, and its own proper fastenings to the bones or other parts which it moves by contracting. In a figure of the human body, which has the skin removed so as to show the flesh, or muscle, we can see the directions in which some of the muscles lie and in which they must contract. In the same way when we look at a piece of meat we see that there is a distinct grain or direction in which the bundles of fibres are laid, so that it can be cut either with the grain or across the grain ; and the fibres lie the lengthway of the muscle.

Many muscular contractions often go to the performance of what we call one movement, and we generally have very little idea of how the obedient muscles carry out the orders we give them to do this or that. So when we stand upright many strong muscles are at work, their contractions being balanced against each other to keep us in this position. One stretching over the front of the knee contracts to keep the knee from bending; the muscles of the calf, the back of the thigh, and up the spine are contracting sufficiently to prevent us from falling forwards, and those up the front of the body are at work to keep us from falling backwards. If the muscle of the back of the neck left off working the head would sink upon the breast, or if the front muscles of the throat failed to contract it would fall helplessly backward ; and we can see that this is just what happens in a swoon, when the usual work is not going on.

When a muscle is at rest it is at full length; all its work, its labour, its effort, is in shortening, and this labour may be either voluntary, as when we move our arms and legs on purpose, or involuntary, like the motions of our heart and our chest during breathing, which go on regularly from our birth till we die, without our ever having to think at all about them.

While we were speaking of bones, we compared the human body only with those of the Vertebrate animals, as they alone have bones, but now the whole animal kingdom can come into comparison, for vigorous muscles are at work when a grasshopper leaps or a serpula spreads its tentacles, as well as in a tiger striking its prey, or a swallow migrating across the sea. The breast muscles of flying birds, by which they move their wings, are very largely developed, and this is the reason why the breast-bone of a bird should be, as it is, extended forward in a sharp keel, since it gives far more space for the attachment of the large breast muscles than it would if it were flat like those of mammals.

Another remarkable development of muscular power may be noticed in the whale, whose vast masses of flesh end in long tendons, which run just like rudder-cords down to the tip of the tail, and by their contractions turn it hither and thither for swimming and for guiding the body.

This body of ours, supported by bones and overlaid with muscles, is hollow, and we want next to know what it contains. Look at this general picture of the contents of the body (see next page), and notice first that it is divided in two by a partition, D, rather above the middle. The partition, which is called the diaphragm, is partly of flesh, partly of a sort of skin or membrane, and is attached by its edge all round to the sides of the body; but as it is not tightly stretched the centre can move up and down. The upper chamber, which is above the diaphragm, is the part of the body enclosed by the bony cage of the ribs and breast-bone, and it contains the heart and lungs. You can see the heart in the middle, just over the diaphragm, D, lying between the large spongy masses of the lungs, which fill up the cavity on each side. The lower chamber, which is soft in front,

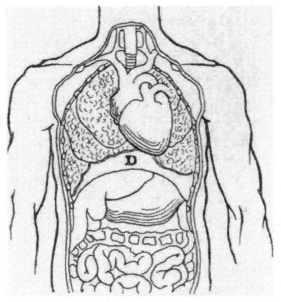

Human Thorax and Abdomen laid open.

is called the abdomen, and contains the stomach, the bowels or intestines, and some other organs.

The heart is a complicated bag, enclosed in a double membrane, and completely divided inside from top to bottom, so that there is no interior communication at all between the two sides of the heart, and the only way in which anything can pass from one side to the other is by going out through the passages which lead out of one side, and going all round to get at the passages which come in at the other—and a very long way round it is. It is like a block of two houses under one roof, each with front and back doors, and no way for the neighbours to meet except by going out at the front door of one house and round to the back door of the

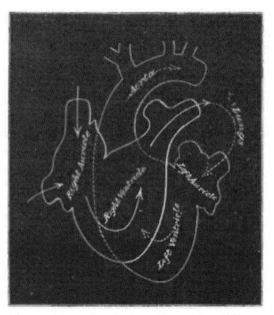

Diagram of the Circulation through the Heart (Dalton).

other. Each division is further divided into two chambers, one above, and one below, making altogether four parts, of which the two upper are called auricles, and the two lower ventricles. The walls of the heart itself are formed of a network of muscles, so that when the muscles contract, they give a squeeze to the heart, squeezing out some of the blood that it contains, just as you squeeze out some water when you tighten your fingers round a damp sponge. The whole of the heart does not contract at the same moment, but the auricles first get a squeeze, and the next instant the ventricles follow, after which there is a moment's rest while they expand again. The blood thus squeezed out runs round all the passages, and comes back into the other side of the heart, from which another squeeze sends it through a second set of passages, not the same as before, round again into the first side; and so, as the heart goes on, every second or oftener, regularly contracting—beating, as we call it—all the blood continually runs round and round, making the double journey and getting back to its starting point in about half a minute.

Let us follow the blood on this journey. Starting from the left ventricle we find that the passage from it is a large tube with firm muscular walls, called an *artery*, which like the stem of a tree branches and branches again and again into ever smaller and smaller tubes, spreading through every part of the flesh of the body. There are arterial tubes passing into the head, the arms, the legs, the different organs of the interior, even into the walls of the heart itself (not into its cavities), and constantly dividing, they come down at last into innumerable very tiny branchlets with very thin walls, which are called capillary tubes. These capillary tubes are literally everywhere; you cannot make a prick in any spot of the body without fetching blood from them—and the blood in them gives the red colour to the flesh. Now the blood, as it travels along the arteries, brings with it all the supplies needed for the nourishment and growth of

155

the tissues of the body, and *through* the very thin walls of the capillaries, as through a filter, every one of these tissues, as the different materials of the body are called, draws from the blood the particular substances it needs; the bony tissue takes what it wants to make bone, the muscular tissue the supplies for muscle; the cartilages, the fat, the skin, each finds what it needs; and, more-over, each of these gives back into the blood to be carried away all that it has used up and done with, consisting chiefly of water, ammonia, and a gas called carbonic acid.

In fact, the blood is like a merchant everywhere selling food to the tissues, and buying up and carting away their dust and rubbish. It does not itself go out of the capillaries, but travels always inside the vessels, carrying on these transactions through the walls, as the huckster might stay always in his van, giving out and taking in goods through the windows. After this distribution the capillaries begin to flow together again, as little rivulets flow into river channels, until all the distributed streams are gathered again into large channels, finally discharging themselves in one great vessel into the *right auricle* of the heart. As the vessels carrying blood to the capillaries were called arteries, and the blood in them arterial blood, so the vessels bringing it back to the heart are called *veins*, and the blood in them venous blood. Instead of being rich bright red, the venous blood, charged with carbonic acid, etc., is of a much darker red colour, and the veins themselves are not so firm and muscular as the arteries; their walls are flabby and fall together when empty. The next contraction of the auricles drives the venous blood forward to the *right ventricle*, and when the ventricles contract in their turn it is again squeezed out into a great artery called the pulmonary artery, which, dividing up as before, carries it this time to the lungs, where it spreads out again into capillaries. Through the walls of these the carbonic acid and some of the water are filtered out into the little spongy holes of the lungs, and when the chest contracts and squeezes the lungs the air is forced up the windpipe and out through the nose and mouth into the atmosphere. As the chest expands again, clean, fresh air is drawn into the lungs, and the blood seizes upon the particles of oxygen gas which it contains, draws them into its tubes, and carries them along, purifying and brightening itself with them all the way, as it travels back through larger and larger veins till it is discharged into the *left auricle* of the heart, from whence it is squeezed into the *left ventricle*, ready to begin its course all over again.

The journey is always made in the same direction, and blood can never flow backwards the wrong way, as the openings of the heart and the passages of the veins are protected by valves, which are like swing doors, only opening one way, letting everything pass freely in the right direction, but shutting tight against any pressure from the wrong side.

The lungs are not the only filter by which the blood is purified—they drain off most of the carbonic acid and a good deal of water; but the skin is also at work getting rid of water in the form of sweat, and the rest of the water, with the other refuse substances, is filtered through the kidneys into the bladder and so passed out of the body.

If we prick our finger, so as to draw a drop of blood, and then examine the blood through a microscope, we shall find that instead of being all a red liquid, it is really an almost colourless liquid, in which numbers of tiny red things, like specks of red jelly, float about and give it its colour. They are round and rather flat in shape, like thick pennies, and as the blood

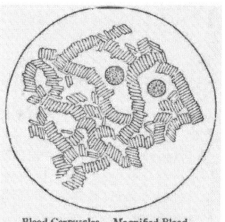

Blood Corpuscles. Magnified Blood.

cools they have a tendency to stick together in rolls, like a roll of copper coins. These are called the red corpuscles —tiny things—three thousand of which would lie side by side in an inch. Among them wander some white corpuscles, which are fewer in number than the red, larger, and not flat; they might rather be called globular, but these wonderful little things are always changing their shape, and, in fact, have a great likeness to the Amœba, which we read of as the last and simplest of the animals. The liquid in which the corpuscles float is called *plasma*, and has the property, when taken from the body, of thickening like the white of egg when boiled, and making what is called a clot.

The blood of all the Mammalia contains two kinds of

corpuscles, like that of man; Birds, Reptiles, Amphibia, and Fishes have red and white corpuscles in their blood, but the red ones are oval in shape, not circular like those of Mammals. Those of the Invertebrate animals which have true blood at all, have only white corpuscles in it. The corpuscles vary a good deal in size in different creatures, and, strange to say, they are largest in the Amphibia.

The whole arrangement of the circulation of blood gets simpler as we look

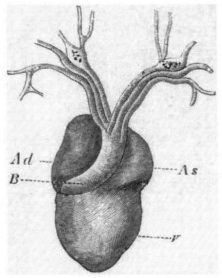

The Heart of a Frog (*Rana Esculente*) from the front.—*V*, ventricle; *Ad*, right auricle; *As*, left auricle; *B*, bulbu arteriosus, dividing into right and left aortæ. (Ecker.)

Heart of a Fish.

down the series of the animals. In the Reptiles the two sides of the heart are not completely separated except in the Crocodiles; and Frogs have two auricles but only one ventricle; Fishes, whose blood is aired through the gills instead of through lungs, have but one auricle and one ventricle. Among the Invertebrates some Mollusks have a simple form of heart with auricle and ventricle. In Arthropods and some worms the heart is an enlarged portion of the main blood-vessel which pulsates.

How is the blood supplied with all the materials which it carries to the tissues of the body? We take them in daily in our food, and the whole use and object of feeding is thus to supply the blood. When we put food into our mouth, it is, or should be, first chewed by the teeth, and softened and somewhat changed by the *saliva*, or

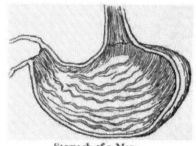

Stomach of a Man.

moisture of the mouth, and then it goes down the throat by a tube which passes through the diaphragm and into the stomach. It does not *fall* down the throat, but the

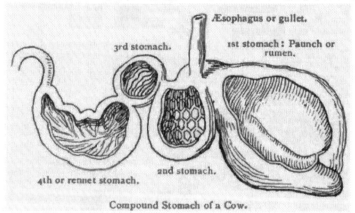

Æsophagus or gullet.

3rd stomach.

1st stomach : Paunch or rumen.

2nd stomach.

4th or rennet stomach.

Compound Stomach of a Cow.

muscular rings surrounding the tube contract in regular succession, pushing the food before them. This enables us to swallow when we are lying down ; and, of course, horses and all grazing animals have to swallow upwards.

The stomach is a sort of bag, holding, in a full grown man, about two or three pints, and supplying a juice called gastric juice, which dissolves or digests the softened food, and makes it into a kind of milky fluid. Ruminating animals have the stomach divided into several compartments, and the one into which the food comes first returns it, as you will remember, into the mouth again, for chewing the cud. When it is swallowed the second time, it passes at once into the third and thence to the fourth stomach, in which alone is found the gastric juice for dissolving it (see picture on last page).

From the stomach the soft pulpy mass goes out into the intestines or bowels—a very long tube coiled up in many irregular coils, and filling all the lower part of the abdomen, p. 148. Here other juices are poured in to act upon the food; the wall of the intestine itself supplies what is called intestinal juice ; pancreatic juice is furnished by an organ called the pancreas, and the liver adds the greenish yellow fluid known as bile. All these juices are selected, or, as we say, secreted, by the different organs out of the materials brought by the blood, and poured either directly into the stomach and intestines, or conveyed there through tubes.

All the substances that we eat for food can be arranged into three divisions: (1) flesh-forming matter (called *proteid* matter) ; or (2) fatty and oily matter ; or (3) starchy and sugary matter, which is furnished by the bread and vegetables of our food ; all the rest is either water, or mineral substances, such as salt. Now, of these three kinds, the flesh-forming matter is dissolved by the gastric juice, the starch by the saliva of the mouth and the pancreatic juice, and the fat is dealt with by the bile and pancreatic juice; so that all the goodness of the food becomes quite liquid, and is drawn into the blood through the busy capillaries which are found crowded in all the walls of the intestines. The great length and many folds of these give time and opportunity for the whole nourishment to be thus taken into the blood, leaving nothing in the tube but the indigestible refuse, or what we might call the papers and packing-cases in which the nourishment was packed up in our food, together with what remains of the dissolving juices.

The whole of the food passage from the mouth downward, including the stomach, is called the alimentary canal, and its muscular rings, contracting one after another, keep pushing the food forwards, until, the goodness being all extracted, the remaining mass is finally pushed out of the body through the bowels.

We must now turn to consider by what means all these processes—the muscular movements, the circulation of the blood, and the digestive apparatus—are regulated. The connecting and controlling organization which governs and directs and keeps everything in order is found in the nervous system, which has its centre in the brain.

The brain, which occupies the large hollow of the skull, consists of soft grey and white nerve matter, arranged in definite parts, and folds, and shapes. The same nerve matter is continued all down the tube in the vertebral column, and is there called the spinal cord; and both the brain and the spinal cord give off many branches or *nerves*, which spread, like the blood-vessels, into every part of

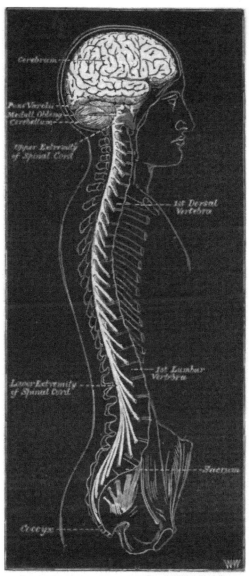

Diagram showing the Human Brain and the Spinal Cord.

the body, in the form of very fine white threads. The nerves that start from the brain, coming through little holes in its bony case, chiefly supply the head and face, while the spinal nerves, issuing between the vertebræ all down the back, spread their branches into every muscle, and into the whole surface of the skin.

Now, these fine white threads act the part of telegraph or telephone wires. Whatever impressions from outside are made upon the body—of injury or resistance, heat or cold, scent, sound, or light—are instantly telegraphed through the nerves to the brain, and there become sensations or feelings. The brain in turn telegraphs orders through nerves to the muscles when they are to contract, and so bring about various movements. The nerves which bring information to the brain are called *sensory nerves;* they end chiefly in the skin, and are not very evenly distributed, some sensitive parts, like the tips of the fingers, being crowded with nerve fibres, while the back of the hand has comparatively few. Those which carry orders from the brain to the muscles are called *motor nerves;* they end in the muscles, and no muscle ever contracts unless it receives a nerve message or stimulus to do so.

It does not at all follow that we are always conscious of the nerve orders given in our own bodies, or that we can always control them. The nerves bring in a message that the skin is cold, back flashes the impulse to shiver; the nerves of the eye say something is approaching or touching the eye, and the eye-lid instantly winks in obedience to an unconscious command.

Nay, sometimes we would give anything to prevent the action. The ear nerves report words that are heard;

why should the muscles controlling the blood-vessels of the face and neck immediately have the orders for their firm contraction revoked, so that they are overfilled with blood, and make us blush scarlet? We even try to resist—endeavouring, perhaps, to keep cool and firm, when something has ordered our heart to throb and our hand to shake.

All these cases of movements without our control, and often without our knowledge, are said to be cases of reflex action, as if messages from the sensory nerves were simply *reflected* back from the nervous centres to the motor nerves; and it is remarkable that to produce reflex action, it is not always necessary that the message should reach the brain itself; for not only can the spinal cord issue orders to the muscles, but it seems that this power is even shared to some extent by a series of little masses of nerve matter, known as the sympathetic ganglia, which lie in front of the vertebral column, and are connected by nervous cords with each other, and the nerves of the brain and spine.

The seat, however, of *sensation*, and of all the mental powers, intelligence, reason, thought, is in the brain or *cerebrum* only; and it alone is the source of *voluntary* muscular contraction.

Why the nerve messages excite feelings in the brain, how the brain is used in reasoning or remembering, or how the living being which exercises control by its means, is connected with it; to these questions science has no answer to give. The mystery of life is above and beyond its reach, even while it examines into the sources of sensation, the machinery for motion, and the carrying out of the orders of the will. That the nerve messages are carried as above described is proved by the facts that when certain nerves are cut or injured, sensations of certain kinds, or of certain parts of the body, are entirely stopped, while injury to others takes away the power of muscular contraction from the limbs to which these nerves extend; that division of the spinal cord puts an end both to feeling and motion in all parts supplied by spinal nerves issuing below the injury; but that thought and intelligence are not affected unless the injury is in the brain itself.

A few words must be said about the special organs of the senses. We *feel* through the nerves all over our body, but we can only taste and smell with the mouth and nose, hear with the ears, and see with the eyes. The special nerve which produces the sensation of smell has its delicate branches spread over the inside of the nose, while its trunk ends in the brain, and the nerve of taste spreads in like manner over the back of the palate or roof of the mouth and the tongue, and before either of these can act they must come into contact with the actual particles of substance which are tasted or smelt. But light and sound are brought to us imperceptibly, without our being in actual contact with the luminous or resounding substances, so that the eye and ear are our means of communication with other men, and with the outer world, while to the eye is also given the wonderful power of searching the unspeakable distances of the starry sky.

The ear is hidden deep within the skull, only the outside porch being visible. It consists of an intricate labyrinth of chambers and winding passages, some of which are filled with air, and some with liquid, entirely closed from outside by fibrous membranes. One of the inner chambers carries on its walls a set of marvellously fine fibres, like microscopic harp-strings, each of which responds to a different pitch of sound; and when any audible vibrations come floating on the air, the membranes, the air, and the liquids of the ear-chambers carry them inwards till they reach and strike the dainty harp-strings, which, being in close connection with the ends of the auditory nerve or nerve of hearing, excite it to telegraph its message of sound into the brain.

Observe also the power of the ear to hear, and distinguish many sounds at once, as in the harmonies of music. " Picture to yourselves the contrast between a great orchestra containing some hundred performers and instruments, and that small music-room built of ivory, no bigger than a cherry stone, which we call an ear, where there is ample accommodation for all of them to play together. The players, indeed, and their instruments are not admitted, but what of that if their music be? Nay, if you only think of it, what we call a musical performance is, after all, but the last rehearsal. The true performance is within the ear's music-room, and each one of us has the whole orchestra to himself." *

The eye, through which the brain receives its messages of light and vision, is a hollow globe of tough tissue (the same material of which tendons are made), to whose outer surface are attached six different muscles enabling the eyeball to be moved up and down or from side to

* "The Five Gateways of Knowledge." George Wilson.

side, so as to see freely round. In front of the globe the tissue is modified into a clear window, something like a watch-glass, called the cornea, which is always kept spotlessly clean by the winking of the eyelids. A little way behind the window is placed the elastic crystalline lens, or magnifying glass, set in a delicate membranous frame, whose edges reaching to the walls of the eyeball, divide the hollow chamber into two parts, both of which are filled with liquid; and close in front

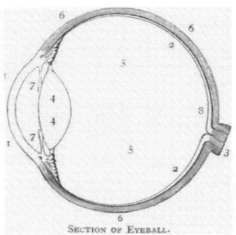

SECTION OF EYEBALL.

1. Cornea; 2. Retina; 3. Optic nerve; 4. Crystalline lens; 5. Vitreous humour; 6. Sclerotic coat; 7. Iris; 8. The yellow spot, the central and most sensitive part of the retina.

of the lens, between it and the cornea, hangs the iris, a coloured curtain with a circular opening which enlarges or contracts, so as to regulate the amount of light passing into the eye. On looking attentively at another person's eye, or at your own in the glass, you can see the iris as the coloured part, in the centre of which appears what we call the dark pupil of the eye, which is really part of the dark hollow seen through the clear crystalline lens. When the light is very bright, the hole in the iris contracts, and the pupil appears small; but in the dusk it enlarges so as to admit more light. If you keep your eyes shut for a few minutes, and then open them suddenly before a looking-glass, you can see the pupil which had become larger in the dark rapidly contracting as the light falls upon it.

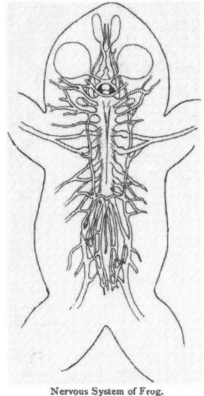

Nervous System of Frog.

The back of the eyeball is lined with a soft dark curtain, on the inner side of which is spread a very delicate white membrane called the retina, containing the fine branches of the optic nerve, or nerve of seeing, whose trunk end is in the brain.

Turning now to the animals, we find that the nervous system of the Vertebrata is on the same pattern as that of man, with its centre in the brain, and with the spinal cord giving off the nerves which spread into fine branches through the body. Here, for example, is a picture of the nervous system of a frog. The intelligence of animals seems to vary partly with the size of the brain, and partly with the number of folds in it; and many of them, as we know, have the special senses keenly developed, as, for instance, the sight in birds of prey, and the scent in dogs or ferrets.

But when we come to the Invertebrate creatures, the highest nervous centres consist of ganglia, sometimes collected together into larger nervous masses, sometimes placed singly, one in each segment of the body, and connected by nerve cords, as is shown in this picture of

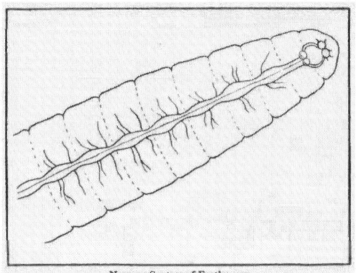

Nervous System of Earthworm.

the nervous system of the common earthworm. Yet in the Insects, for instance, such a large and important nerve-mass is present in the head that it is often called the "brain." Still lower in the scale, as in the starfishes, even ganglia disappear, and we find a mere ring of nerve round the mouth, with cords radiating from it into the principal divisions of the animal.

At the same time the eyes, which in insects are

elaborately compound, and in spiders numerous, become in lower forms mere sensitive points, perhaps dimly conscious of light, but not perceiving shapes or colours; until, finally, we come down to creatures of the simplest organization, without perceptible nerves, circulation, or even stomach, and hardly to be distinguished from the lower forms of plant-life. Indeed, in these borderlands of animal and vegetable nature, the main difference seems to be that, while plants can take in and digest inorganic substances, thus converting them into organized matter, animals can only sustain life upon food already organized either in plants, or in other animals.

CHAPTER IX.

PLANTS.

PLANTS have, for our present purpose, the very great advantage over animals that they stand still to be looked at, and, as they are to be found almost everywhere, you are now invited to examine and see their arrangements for your-selves, and not only to read about them, as we are often obliged to do with animals.

Buttercup Plant.

First, then, let us go and dig up a Buttercup plant by the roots, the largest we can find, and, shaking it as free as pos-sible from earth, lay it on the table to look at. At the bottom come the roots, a mass of fibres, which were all buried out of sight underground; above these shoots up the strong green stem with many branches, carrying a number of green leaves, and at the top are the large, bright, shining, yellow flowers. If some of the flowers are already over, we may see their places occupied by the seed-vessels, which were inside the flowers, and which contain the seeds.

Now, if we look at an Oak tree we shall find that it

Oak Tree.

also has a mass of roots underground, a strong stem shooting up, with branches carrying many green leaves and flowers, within which are the seed-vessels containing the seeds. And, in fact, this is the general character of all flowering plants, though there are many variations in details, some of which we will look at.

We shall now need some more plants for comparison, and had better dig up a Daisy, a Plantain, a Dandelion, a Potato, a Turnip, a Crocus, if we can find one, and a Lily of the Valley; but these two last bloom in the spring, and their flowers will be over when the others come out. Lay them in a row, and look at the roots of each. The roots of the Daisy are all fine fibres, starting immediately from under the leaves, but the Plantain (p. 170) has one thick long root growing straight down, out of which come most of the fibres ; this is called a tap-root. The thick hard part of the Dandelion root is a tap-root growing downwards, but its upper part lies near the surface of the ground, and carries buds, from which fresh stems will rise; it is, in fact, considered an underground part of a stem—not a true root : its name is a root-stock. In the Potato (p. 171) we find hard fleshy masses, containing several eyes, or buds, out of which the next year's plants would grow ; these are, in fact, potatoes, and are swollen stems or tubers. The Turnip root spreads out into a large round head, a great part of which pushes up, and appears above ground, while the fibres hang down from it below; this is an enlargement of the root itself, the globular root. The bulb of the Crocus is a true root-stock, having buds concealed under scales. The root-stocks of the Lily of the

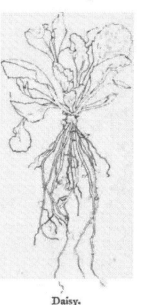

Daisy.

Valley (p. 171) spread sideways underground, and throw up the fresh growths from their buds.

We thus see there exist great variations in the forms of roots, fibres only, which may be quite simple, like hairs, or branched, as in grass-roots, or large and woody, as in the roots of trees, tap-roots.

For stems and branches, let us add to our collection

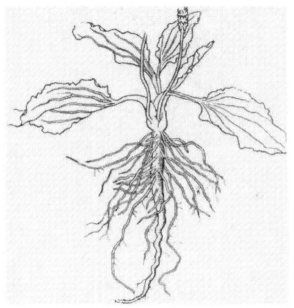

Plantain.

some Bind-weed or garden Convolvulus, Bryony, Virginia Creeper, and a Rose branch. But first we must step outside, and look at some trees which we cannot bring in. On p. 172 is a young Spruce Fir which would make a capital Christmas-tree. Its central stem is as straight as an arrow from foot to tip, and the branches are thrown out from its sides regularly all round. How different it is

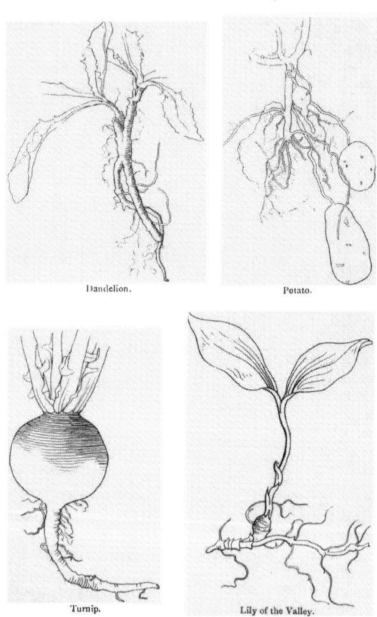

Dandelion.

Potato.

Turnip.

Lily of the Valley.

from an Oak, which has no special leading shoot in the centre, but whose branches divide up continually into smaller and smaller forks and twigs, until they make a

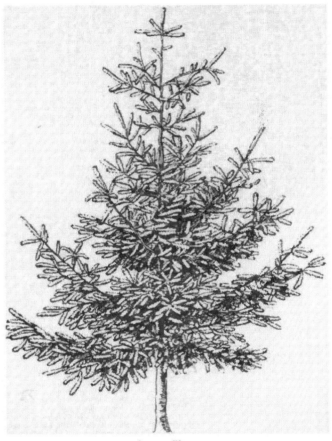

Spruce Fir.

sort of round head to the tree. Both these forms of growth are common among plants large and small. The Buttercup, you see, branches like the Oak tree, but Mullein and Hollyhock plants grow up with side-branches

177

round a leading shoot. The Buttercup and Mullein stems are strong enough to stand upright, but our three next specimens are not, and take different ways of helping themselves up. The Convolvulus trails on the ground until its stem touches some support, when it twines itself

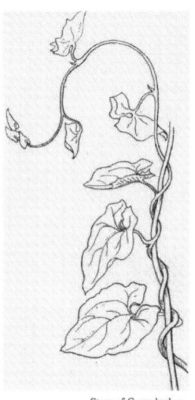 

Stem of Convolvulus.   Virginia Creeper.

round and runs up it, growing spirally. The Bryony or Pea (p. 174) puts out from the side of its leaf short green threads called tendrils, which curl round any support they can find, and so hold up the main stem; but the tendrils of the Virginia Creeper may behave quite differently. Instead of twisting round a support, they grow straight out till they touch a wall, or something to which they can attach a little sucker at their tip. As soon as this is fast, the tendril shortens itself up into a tight coil, so drawing the main stem close to the wall. The Rose prickles also help the plant to climb, as it thrusts up its slender shoots through other bushes.

Leaves have endless varieties of shape, and perhaps a walk through the kitchen garden will serve as well as anything to show a number of them. Look at the large, juicy leaves of the Cabbage, folded one outside another over the heart; the long, slender leaves of the Onions; the handsome, heart-shaped leaves of the Scarlet-runners; the notched leaves of Strawberries; the soft, plumy leaves of Carrots; and the light, leaf-like feathers of Asparagus. But, whether large or small, pointed or round, thick or thin, whole or cut, these all have one thing in common, and that is their green colour, which is, as we shall see presently, of the greatest

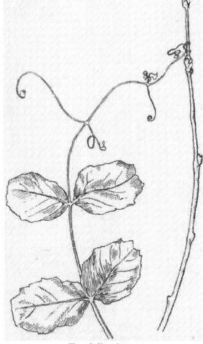

Tendrils of Pea.

importance to the health and usefulness of the plants.

In plants that send up a main stem and branches, the leaves usually grow upon these, sometimes sitting close to the branch, like the upper leaves of Honeysuckle, or more often joined to it by a leafstalk (called a *petiole*); but many plants have no obvious main stem. Look at the leaves of a Daisy (p. 169); they lie in a flat rosette upon the ground, each one growing directly from the root-stock, from which the flower stalks also rise. Strawberry leaves mostly grow straight out of the root-stock, but, as they have long leafstalks, they make a tuft and not a rosette. Some plants have both root-leaves and stem-leaves, like the great Ox-eye daisy, whose root-leaves are broad and set on long stalks, while the stem-leaves are narrow and have no stalks. A different arrangement is shown in the leaves of Grass, the lower part of which is wrapped round the grass stem like a sheath.

Daffodils (p. 187), Hyacinths, Lilies, and many others send up their flower stems in the same way from the centre of a sheath of leaves.

Now we come to the flowers themselves, and the arrangements in which they grow, of which we can find plenty of varieties in the fields. The bright scarlet Poppy carries its flowers singly, one on the end of each main stem (p. 176), while the little red Pimpernel also has separate single blossoms; but instead of being at the end of stems, their flower stalks each grow out where a leaf joins the stem—in the *axil* of the leaf, as we say. The flowers of the Plantain (p. 176) are crowded closely together all along a spike, and another kind of spike may be seen in the catkins that hang on the Willow trees. We might call the flowers of the Cuckoo-flower a spike, too ; but as each blossom is set on a flower stalk, or *pedicel*, the

Flower of Poppy.   Plantain.

whole arrangement is much freer and looser, and its proper name is a *raceme*. A *panicle* is like a raceme in

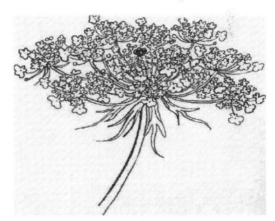

Cuckoo-flower.   Carrot.

which the flower pedicels are themselves branched so as to carry more than one flower each.

The flowers of Carrots, Parsley, and Hemlock show quite another pattern. Their flower pedicels all start from the same place, but the outer ones are longer than the inner, so that the top of the flower head is flat. This arrangement is called an *umbel*.

If we pull a daisy to pieces, and, separating some of the little yellow things from its centre, examine them with a magnifying glass, we may be surprised to find that each is a complete little yellow blossom by itself; and so, also, though of a different shape, are the little white rays which surround the yellow disk. So, then, a daisy, instead of being one single blossom, is a whole crowd of tiny blossoms, some yellow and some white, set in a regular order, and making up together a head. A great many flowers are arranged on this general pattern — all the thistles and hard-heads, and ragworts and groundsels, the chrysanthemums and asters and dandelions—and many others.

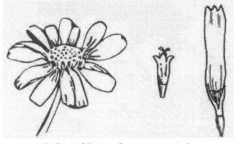

Daisy, with two florets separated.

To see the parts of flowers we had better take a single perfect blossom, and examine it carefully. An Apple blossom will do well to begin with, and we will cut two, as we shall need to pull one to pieces. Take one by the stalk, turn it upside down, and look first at the outside; you will find that its outermost covering consists of five green pieces, something like thick green leaves (p. 179). This green covering is called the *calyx*, and each piece when separate is a *sepal;* it is a calyx of five united sepals. Within the calyx come five large, thin, delicate, white or pinkish leaves, which are called the *corolla;* each separate leaf is a *petal,* and the calyx and the corolla together form the *perianth.* Now carefully pick off all the petals, and see what is left inside. The next thing is a number of slender, thread-like stems, each carrying a small head. The

Apple Blossom.

name for them is *stamens,* the heads are called *anthers,* and the stems *filaments.* Cut them all away also, and we have left five thicker green stems joined together at their base. We must see what is at the bottom of these; so, with a sharp pen-knife, cut straight down the middle, laying open the centre of the flower as in the illustration, and we find a seed-vessel, or *ovary*, with five divisions, out of the top of which grow the five green stems. The stems are called *styles*, their swollen tips are *stigmas*, the divisions below con-taining the seeds are *car-pels*, and the whole organ is called the *pistil*. It is pro-perly described as a pistil of five carpels, with five styles joined only at their base.

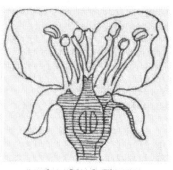

The four series here de-scribed—the calyx, corolla, stamens, and pistil — are called the four *floral whorls*.

Section of Apple Blossom.

A flower is said to be *complete*, when, like the apple, it has them all four; and *regular*, when each of the sepals, and each of the petals, is like the rest of the set.

Compare the buttercup with the apple blossom. It has also five green sepals, either spreading in the same way under the petals, or else folded back towards the stem. Leave these; but pull off the five (or more) bright, golden petals, and also the numerous stamens, until there is only the green pistil, which we will cut open down the centre

Section of Pistil and Recep-tacle of Buttercup.

as before. This time we find quite a different state of things. Instead of a seed-vessel divided into five carpels, there is a raised, green pyramid, upon which sit many carpels quite separate from one another and without styles, but ending in a sort of beak, at the point of which is the stigma.

In this case the ovary, or set of seed-bearing vessels, is said to be *superior*, or set up above the calyx; whereas, in the apple, it is *inferior*, or sunk into the swollen top of the flower stalk, which is so continuous with the calyx that we cannot say where one ends and the other begins.

In the buttercup the green pyramid is the enlarged tip of the flower stalk. When this rises above the

calyx, it is called the *receptacle*, and takes different forms in different flowers, sometimes making a flat cushion or disk, and sometimes reduced to a mere ring.

The calyx and the corolla are very different in different plants, both in their colour, in the number of the sepals and petals, and in their arrangement.

The commonest variety in regular flowers consists in their being joined together for some part of their length.

Primrose Blossom.

Thus in the Primrose the lower part of all the petals is joined into a tube, while the upper parts are widely spread. The sepals of the calyx are joined in the same way almost to the top, and form an outer tube.

It is not uncommon to find plants which instead of having a double perianth, that is, both calyx and carolla, have but a single whorl in their place; and as it is difficult to see which represents the calyx or the corolla, it is generally spoken of simply as the perianth. The Lily of the Valley is one of these flowers of a single perianth which is united into a bell shape, cleft at the outer edge into six points.

In irregular flowers the variety of shapes is endless. Take a blossom of Sweet-pea, or Bean, or Laburnum, or Broom, which are all alike, and count its petals. There are five, of different sizes. The two smallest, often joined together by their edges into a boat shape, are

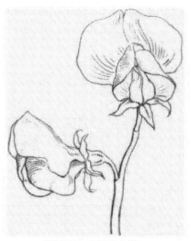

Lily of the Valley Blossom.

Blossom of Sweet Pea.

also the innermost; they are known as the keel. On each side stand the two wings, and over and outside all is the standard, as it is called, which in the bud was folded over all the rest. If you carefully pull off each of the petals you will be able to see that enclosed in the keel is a bundle of ten stamens, nine of them being always joined together, while one stands apart, and also a pistil, the lower part of which consists of a single long narrow carpel, which will presently develop into the pod.

For some other irregular forms, compare for yourself the blossoms of Snapdragon, Dead-nettle, Violet, Orchis, and Honeysuckle.

Now and then a flower has no perianth at all. This is the case with the tiny blossoms crowded upon a catkin, and with the flowers of many sedges.

If flowers can thus dispense with a perianth altogether, it is clear that this, though the most showy and attractive to insects, is not the most essential part. The real importance of the flower lies in the pistil and stamens, and their action upon each other is as follows.

Within the carpels of the pistil are the tiny *ovules* which are to become the future seeds. They do not lie about loose like stones in a bag, but are attached to the inside of the carpel. Ovules, however, cannot develop into seeds, unless they are fertilized by the contents of the anthers, or heads of the stamens. When the anthers are

Ovules of Pea in Pod.

ripe, they open and thus let out fine grains of powder, called pollen, which is usually yellow. The pollen grains are tiny cells, and when one reaches the spongy, sticky stigma of the pistil, soon there grows out of it a tube which passes into the stigma, and down the length of the style into the ovary, where it communicates its contents into the ovule. It is a beautiful sight to see through a microscope the pollen grains putting out their delicate tubes.

The stamens and pistils do not always grow in the same flower. On a hazel bush we shall find that all the hanging catkins contain stamens only; but by careful search over the tree, minute crimson tufts may be found, which are the pistil flowers and will hereafter produce the nuts. In willows the separation goes still further, and the stamen and pistil flowers—the male and female catkins—grow on different trees. Even where stamens and pistils occur in the same flower, the seeds are found to be finer when fertilized by pollen from flowers at a distance, and you will be able to read in larger books of many exquisitely beautiful natural arrangements for securing fertilization from distant flowers, of which there is not room to speak here. Thus at last appears the seed, the plant's treasure, which, born in the heart of the flower, protected by all its envelopes, nourished, and ripened, at last escapes, either falling to the ground, or carried away by wind, or water, or birds, to grow into a new plant.

To watch the beginning of this new growth, we cannot choose a better seed than a bean, which is large enough to be well seen, and also has the advantage of being very familiar. We all know the appearance of beans when cooked for dinner—with a loose grey skin, inside of which are the two fat, fleshy, green halves of the bean, joined, if still joined at all, only in one spot of their edge. If we take off the skin of a raw bean and observe this joining closely, we shall see that it is made by a tiny thing, with one end more pointed than the other, which lies between the two halves of the bean, and is attached by its middle to both of them. When a bean is sown, that is, placed on or just under the soil where air can reach it, and furnished with moisture and sufficient warmth, it will begin to sprout, as we say; that is, this tiny body will begin to lengthen at both ends. The pointed end will break through the skin of the seed and appear outside as a little tail, which is called the *radicle*, and is the beginning of the root of the plantlet; while the other end will grow in the opposite direction up between the two halves of the seed, and soon appear

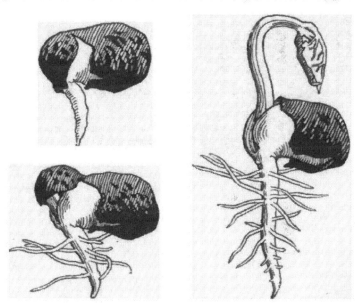

Germination of Bean.

above them; it is called the *plumule*, and is the beginning of the stem and leaves. The fleshy parts of the seed are called the *cotyledons*, and they are so fat because they are filled with the nourishment which the young plant requires at first. As the plant grows on and absorbs this nourishment, the cotyledons shrivel up and die.

This history is not quite the same for all plants. In

the mustard seed, for example, the cotyledons are small and thin, and instead of remaining on the ground feeding the plantlet, they grow up above ground below the growing plumule, spread out, turn green, and become the first leaves of the plant, drawing in nourishment for it from the air. We should never have guessed from the bean that the cotyledons were the first leaves of a plant, had we not clearly seen it in other plants. Some seeds contain, besides the young plant with its cotyledons, a separate store of nourishment.

Mustard Plant, with Cotyledons and Second Leaves.

These seed leaves are not generally like the later leaves of the plant, but the young plant continuing to grow up between them begins to put out its characteristic leaves with the very next pair. We often, therefore, cannot recognize what a seedling will prove to be until its later leaves have appeared.

All Flowering Plants are arranged in three main groups, and we find that the number of the cotyledons is one of the distinguishing marks by which they are recognized.

I. *Dicotyledons.*—The bean and the mustard, as we have seen, each have two cotyledons, as have also far the larger number of the flowering plants, and all our English trees except the fir-tree group. From this circumstance the name Dicotyledons is given to the Class. All the plants of this Class have their branches and twigs growing from buds which sprout out of the sides of the stems, and the parts of their flowers, sepals, petals, stamens, etc., are generally arranged in fours, fives, and eights. If you examine wallflowers and primroses, for instance, you will find that the wallflowers have four sepals in the calyx and four petals in the corolla, while

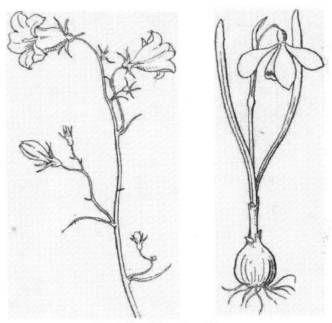

Harebell and Snowdrop Plants.

primroses have five of each. The Class contains many Orders, with flowers both regular and irregular. Two of the largest Orders and most marked types are the Leguminosæ, or Pod-bearing plants, of which the Pea-flower (p. 181) is a representative, and the Composite plants, like the Daisy (p. 177), consisting of many small florets crowded upon one receptacle.

II. *Monocotyledons.*—Crocuses and Snowdrops do not put up a pair of little leaves like the Mustard, but begin with a single cotyledon, and so belong to the Class of Monocotyledons. These rarely have true branches, but their fresh shoots, instead of budding out of the side of the stem, spring out of the middle or heart of the plant, and the leaves or leaf-stems

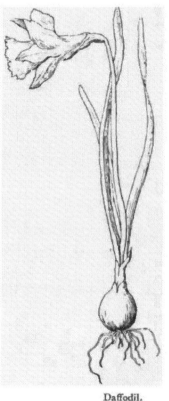

Wallflower.   Daffodil.

frequently form a sheath round the flower stem. The parts of the flowers are generally in threes or sixes.

Here are pictures of a Harebell and a Snowdrop, which show characteristic differences between the two Classes. The harebell, which belongs to the Dicotyledons, has both calyx and corolla divided into five parts, and contains five stamens, and the little plant grows up with slender branches coming out of the sides of the stems. The snowdrop, on the other hand, has three outer sepals of pure white, three short inner ones tipped with green, and six stamens, and the flower-stem grows up in the innermost heart of

Palm Tree.   Germination of Spruce Fir.

the plant, the lower parts of the leaves wrapping it round like a sheath.

This Class includes most of the plants raised from bulbs, such as Hyacinths, Tulips, Daffodils, Lilies, and also Orchises, Arums, Rushes, Reeds, and the important Order of Grasses, which not only feed our cattle, but afford, in the form of Wheat, Barley, Rice, Maize, etc., a great part of our own food supply. A characteristic *tree* of the Class is the palm tree, which carries a single tuft of great leaves at the top of the generally unbranched stem.

III. *Coniferæ.*—The Cone-bearing trees are a single Order containing all the Fir or Pine trees, Cypress, Cedar, Larch, Juniper, Yew, etc. Baby fir trees, some of the most fascinating of young growths, instead of being content with two cotyledons, often put out a whole ring of them, and the Coniferæ are also distinguished by their ovules and seeds being quite naked, instead of being enclosed in an ovary or seed-vessel, like those of both the other Classes.

Flowerless Plants.—The whole of what has been said as yet applies entirely to Flowering Plants, which constitute the highest division of the Vegetable Kingdom. The Flowerless Plants, which have no true stamens or pistils, and therefore do not form seeds in the same way at all, are a more difficult and advanced study, and we must here be content with learning that they consist of Club-mosses, Horse-tails, curious green plants which in wet places often grow to a considerable size ; Ferns, Mosses, Liverworts, which spread their filmy leaves among moss on damp lawns ; Lichens, which sometimes grow as grey tufts on the bark of trees, sometimes form cloudy patches of various colours on old stones and rocks and roofs ; Funguses, including not only mushrooms and the many plants that we call toadstools, but all the various growths to which are given the names of mould, yeast, dry rot, smut and rust in wheat, and others ; Algæ, to which belong all sea-weeds and some fresh-water weeds; and, finally, Bacteria, minute microscopic fungal plants, which grow and swarm in living and dead animals and plants, and are the cause of many diseases, as well as being, in many cases, very useful scavengers.

CHAPTER X.

PLANT LIFE.

ALTHOUGH there seems to be a likeness between some of
the clearly lower forms, yet, on the whole, life in plants is
very different from that of animals. Plants have neither
heart, brain, nerves, nor, in most, power of moving about;
and yet life of some sort there certainly is, for plants are
born, feed, digest, grow, bring forth young, become sick,
and then part with their lives and die. They, and they
only, can feed directly upon the earth and the air, and
in their bodies the nourishment is changed into food
which animals can digest and live upon; so that without
plants all animal life would speedily come to an end,
for, even those creatures which eat other animals would
find that their prey all died of starvation.

One of the mysteries of plant life is the difficulty of
knowing in what a single plant—a single life—consists.
We may have a plant grown from a single seed, and we
feel inclined to say, "This, at least, is *one*." But a
gardener may take our plant and cut it into a score of
pieces, every one of which when planted will grow roots
for itself, and carry on all the business of plant life. Are
they many plants, or one? For they all grew from a
single seed.

We may go on thus propagating a plant by cuttings for a long time, but there is no real fresh spring of life in cuttings; they are but divisions of the old life, and at last it comes to an end: the strain wears out, as we say, and dies. The only beginning of real new life in plants is the seed, the young one born of older parents, and starting afresh.

Duration of Life.—Plants differ very much in the length of their life. Some spring up from seed, grow to their full size, blossom, and bear seed once, and then die, all within a single season. Many of our garden flowers are among these plants of a year, or annuals, as they are called. Sweet-peas, for instance, have to be sown afresh every year, for when autumn comes, and their blooming is over, they die of old age, and their dead bodies decay and rot into the ground, leaving nothing behind them but the baby seeds which are to grow into the next generation.

Some, such as turnips, radishes, etc., take two years to grow to maturity and do their seed-bearing; and others, like the banana, may grow on for many years before they blossom and bring forth seed; but when this one event of their lives is over, they also often die.

Finally, we come to the very numerous plants which are not exhausted to death by their first seeding, but are able to go on year by year putting forth flowers and seeds. In this case they either die down to the root-stock every year, springing up again with new vigour, like the daisy, or else the stem and branches remain, throwing out fresh buds in spring time, like the oak.

When the old leaves hang on the tree till after the

young ones have appeared, as in holly, laurel, and most fir trees, they are called evergreens; but many, as we know, lose their leaves in autumn, and stand all the winter, looking bare and dead, waiting for the next season's growth.

Tissues of Plants.—If we examine the tissues or substance of plants with a microscope, we shall find that

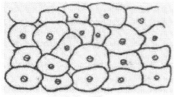

Cellular Tissue.

Woody Tissue.

they are made up of little cells, each of which consists of a bag or case called the cell-wall, and of the contents of the cell, which are generally half liquid in living cells. The cells can sometimes be easily divided from each other, like those in the pulp of an orange, where they are large and distinct, but they are often crowded closely and compactly together.

Generally the cells are round, or

Vascular Tissue.

nearly so, and when living each cell has a small quantity of the living substance, called *protoplasm*, enclosed by the cell-wall. The contents of the cells are not completely separated by this thin wall, but are connected by very fine threads of protoplasm running through the walls. When cells become old they often lose their protoplasm and so become dead, the wall changing from its original *cellulose* to a woody character; the walls at the same time become more or less thickened and sculptured. If now the end walls break down, then long tubes, called *vessels*, are formed, and these, though dead, are not useless, for it is through their cavities all the water needed by the plant ascends from the soil to the leaves. These vessels are also found in the veins of leaves, and multitudes of such vessels, along with woody fibre, give strength to the trunk of a tree.

Feeding of Plants.—This is the way in which plants feed. The delicate cells of the rootlets finding in the

ground the mineral food that they want dissolved in water, draw it through the walls into their cells, from which it passes on, drawn up from cell to cell in the same manner, through the stem and branches until it reaches the leaves. As yet it is almost in the same condition as when it was taken from the soil; but green leaves act as the stomach of the plant and digest

Magnified Tip of Rootlet with Root Hairs.

the food, converting it into various substances, after which it passes down again and is distributed through the cells as is needed. The healthy and vigorous leaves also absorb from the air carbonic acid gas, and separating the carbon and oxygen of which it is made, return some of the oxygen into the air, and use the carbon for nourishment.

The faded and decaying parts of plants, on the other hand, give off a little carbonic acid, and so also do the blossoms, and the green leaves at night; but the quantity is so small as to be absolutely insignificant compared with the oxygen supplied by growing leaves. As, however, the leaves only give out oxygen in the light, it follows that at night or in the dark there is nothing to set against the carbonic acid. Animals, you will re-member, also breathe out carbonic acid and take in oxygen, so that a person who sleeps at night in a small room crowded with plants will have a little less fresh air to breathe, because the plants, as well as himself, are using the oxygen and helping to fill the room with carbonic acid. By day, however, and especially in strong sunlight, just the reverse is the case; the leaves are then rapidly absorbing carbonic acid and throwing out so much oxygen that the plants are helping to purify the air and make it fit for animals to breathe over again. The moral is—keep growing plants in your sitting-rooms, but do not crowd them into your bedrooms.

Take two glass jars or bottles filled to the brim with spring water, and after plunging into each of them a handful of a fresh water-weed, turn them each care-fully mouth downwards into a basin of water, and then put one in the sunshine and the other into a dark cellar. If we examine them both at the end of an hour or two we shall find no change in that brought from the cellar, the leaves having done no work in the dark; but in that which has had the sun upon it we shall see numerous small bubbles formed all over the leaves and also at the top of the bottle, which, if collected after a time and tested, will prove to be pure oxygen gas. In this case the leaves have not indeed drawn their carbonic acid from the air; but they have found some dissolved in the water, and have thrown off the oxygen in it, appropriating the carbon as food.

The principal substance into which the food of the plant, whether procured from air, water, or the soil, is digested in the leaves, is starch, or sugar, which are forms of the same material. The cell walls are made of the starch-like body, cellulose, plenty being wanted for building fresh cells when the plant is growing, while the starch, in little solid grains, is found stored up, either in roots or tubers, like the potato, from which the plant means to nourish itself when it starts growing next year, or in seeds, where it is deposited for the food of the young plants, and from which we take it for our food in the shape of wheat, rice, beans, peas, etc. The sugar, which is the form of the material that can be dissolved in water, occurs, as we know, in the cells of sugar-cane and of sweet ripe fruits.

But different plants, each according to their own powers, secrete out of their food many other substances than these. Some make wax or oils, such as linseed oil, castor oil, olive oil, and others; some produce resin and turpentine, some find lovely colouring materials or sweet scents for their flowers, some give valuable medicines, such as quinine, others deadly poison, like strychnine.

All these vegetable products are deposited in the plant cells, and sometimes there are also deposits made of mineral substances. In many trees, for instance, the cells in the middle of the trunk and old branches become gradually solid, filled up with mineral matter, and unable to do their cell work any more. But a fresh ring of wood grows every year on the outside of these, enlarging the size of the tree, and only through the outer living part are the processes of life carried on : so that an old tree is something like a coral reef, in which the new living coral is growing upon a foundation of older dead coral. Indeed, we know that trees quite hollow inside can go on growing and putting out leaves if the outside rings of sap wood are left. The solid timber of the tree centre is called heart wood, while

Timber cut across, showing rings, and crack during drying.

the younger outside part is sap wood. When a tree is cut down, we can often see on the cut surface of the trunk the successive rings of wood, and, as one is generally made every year, we can, by counting them, tell fairly the age of the tree.

Only green leaves or green parts of plants can digest unorganized matter or the salts of the earth. What happens then to plants which have no green parts? They have to live, just as we do, on *organized* food, some growing as parasites on living plants, either, like the common dodder of our heaths, drinking in the sap of their stems and branches, or else, like broomrape and many others, getting nourishment from

Dodder on heather.

their roots; others feeding on dead and decaying

organized matter, like the mould plant and many other species of fungus.

A few plants vary their diet by catching small insects 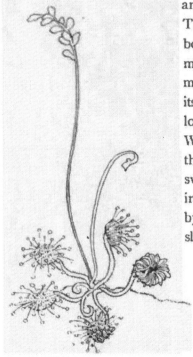 and sucking their juices. The sundew that grows in boggy ground is one of the most familiar of these, and may be easily known by its leaves covered with long red sticky tentacles. When an insect alights on the leaf, attracted by the sweet juice secreted, it irritates and is caught by these tentacles which slowly bend towards it and close over it, at the same time pouring out a fluid, which, like gastric juice, digests the insect, when the nourishment it con-

Sundew plant, and one leaf closing over fly.

tains is absorbed into the substance of the plant. If offered small pieces of raw meat the sundew will accept and digest them in the same manner.

Part II.

THE FORCES OF NATURE.

———

CHAPTER XI.

FORCES AT WORK.

WHEN we look round the world in which we live, we immediately recognize that other things move besides those that are alive. We are quite accustomed to many of these movements; when they always take place under the same circumstances we come to reckon upon them and use them. Babies learn them as they begin to notice what goes on, and by the time we are old enough to think for ourselves we treat them as matters of course, and say, " We know them; they are matters of common experience." Perhaps we are even so mistaken as to think we know all about them.

Now, let us try to observe and study a little more closely.

Gravitation.—Suppose we have a picture standing on a smooth narrow shelf and leaning against the wall. It is not unlikely that, coming one day into the room, we may find it on the floor.

What made it fall? Some one begins to explain to

us that it had not a sufficiently safe support to prevent it from falling. Yes; but why does it want preventing? Does it like the floor better than the shelf? It is not alive; what made it begin moving at all? Why do things tumble down when they are not prevented? These questions seem to cause great amusement, and we are told, "Why, of course they do; they can't help it." Why can't they help it? "We don't know why; but they always do."

No one who had studied the subject for a lifetime could possibly make a better or more scientific answer. *They always do, and we do not know why.* There certainly is a tendency in things to come to the ground —that is, to get as near as possible to the earth, or rather to the centre of the earth, for they will not rest at its surface if they have a chance of getting lower down into it.

Can we find other movements to compare with this inexplicable habit of tumbling down? If we turn from the earth to the heavens we find a precisely similar state of things on a vast scale. For we learn from astronomers that one of the main influences on the movements of the stars is a tendency that they have to approach each other. They seem to have an attraction for each other.

In the same way two balls suspended near each other tend to approach each other, and, in fact, careful observation and experiment show that all material substances have the same tendency, and will come together if they can. Two objects of the same weight, if free to move, travel towards each other at the same pace, and so would meet at a point halfway between them; if, however, one is much heavier than the other, its movements are slower in proportion, and it would not have gone so far before meeting with the lighter one which was hastening towards it. The moon is lighter than the earth, and is always trying to meet it. They do not rush together, nor do the earth and the sun, because another force is present preventing this.

Returning to our fallen picture, we can now see that if the picture and the earth were trying to approach each other, then the instant the obstacle of the shelf was got over, the little picture hurried to the earth, while the huge earth had made such a very small movement before they met as to be quite inappreciable. Moreover, the earth being enormously larger than any of the movable things upon its surface, we are never able to recognize its movements to meet them, and so can only see that the things have a tendency to "come to the ground"—to "tumble down."

Now, men have noticed this tendency of things to come to each other, and though the cause of it is unknown to us, they have given this unknown cause a name. They call it the Force of Gravitation. They have not explained it by giving it a name. Note clearly that we know no more than before about its nature, or how it acts, or why the presence of one body should make a difference to another that is at a distance from it. We can observe further facts about the tendency, and the conditions under which it acts; for instance, that it is not equally strong in objects of the same size, if their weights differ, and that it is stronger at short distances than at long ones, and we are able to set down in figures how the strength and distance vary. But all this is not explaining what the force is, nor the reason that it acts as it does.

Gravitation, the endeavour of things to come to each other, is not the only source of energy in natural objects, else everything would, sooner or later, be stuck together in a vast heap. Other tendencies there are, some of which resist gravitation, and help things to keep apart; we see with great interest that such is the case, but of the causes of any one of them we know very little indeed.

But it is quite time to pick up our picture from the ground. They have got together; they are, so far, satisfied for the time with regard to their gravitation, their mutual attraction. You pick up the picture. By your muscular power you resist and overcome the gravitation, and separate the picture from the ground, but the weight of it upon your arms is the measure of its resistance to being lifted up; that is, of the attraction which the earth has for it. You hang it up by an iron wire to a hook on the wall, and the wall, the hook, and the wire resist the gravitation, and keep the picture in its place on the wall in spite of the constant and steady pull towards the earth.

Cohesion.—Here, then, we have found another force in the wire. It is the force that holds its particles, or molecules (as they are called), firmly together, and prevents them from being pulled apart by the picture's weight. It is called the Force of Cohesion, or molecular attraction. The minute molecules of the wire are so strongly attracted to each other, and cling together so firmly, that they cannot easily be separated.

Cohesion is not the same thing as gravitation, for it acts only when the molecules are very close together indeed. Suppose you cut through the wire, and then press its cut ends closely together again. They will not cohere ; you cannot get the molecules near enough. The moment you leave go, down will go the picture, for the cohesion which held the wire together is interrupted, and the gravitation, no longer resisted, seizes upon the picture, and carries it down to the ground.

Here we have seen three powers opposing one another : first gravitation, then cohesion, then vital power, or the power in the muscles of a living creature, which first resisted the gravitation by picking up the picture, then by hanging it on a wire employed cohesion to continue the resistance, and afterwards, destroying its own work, broke the cohesion by cutting the wire, and let the gravitation have its way again.

Heat.—There are other ways besides cutting in which the cohesion of the wire may be overcome. If the wire were exposed to very great heat, some part of it might melt, and the gravitation, ever active, would instantly get the picture down to the ground. Heat is the agency called in this time, and heat and cohesion are old enemies, always battling together, as we shall have occasion to see again presently.

Chemical Affinity.—Or the wire may escape both cutting and fire, and hang on for years in the same spot. The picture remains in its quiet corner. Is it at rest ? Apparently. The gravitation towards the ground is still

as ready as ever, but the cohesion of the wire opposes
it, and rest in the world of nature only means that
opposing forces are evenly balanced, so that neither
can overcome the other. People go about their business
and their pleasure, and change, and grow old; and if
any heed is sometimes taken of the old picture in the
corner, no one takes heed of the old wire. But at last,
some day or some night, down comes the picture with
a crash. The ever-active gravitation has won; what has
weakened the cohesion? Some one picks it up and
says, " Why, the old wire is eaten through with rust."

Here you are to be introduced to a new agency.

You must know that iron and oxygen have a strong
liking for each other, and that they will come together
when they get the chance; very slowly if the oxygen is
contained in dry air, more rapidly if there is moisture
as well. But the iron and oxygen are not satisfied
merely to be near each other as if they were acted upon
by gravitation; nor do their molecules simply cohere
together, each remaining what it was before. No; they
combine together to form a new substance, which is
neither iron nor oxygen, though it is made of them both;
it is called oxide of iron, or, more familiarly, rust. It
clearly has not the same properties as iron, since, for
one thing, its cohesion is much less strong, and when
much of the iron is converted into rust, the gravitation
does not meet with enough resistance, and wins the
day.

This tendency of substances to combine together and
so form new substances is called Chemical Attraction
or Chemical Affinity, and the examination of these

combinations, and of the conditions under which they take place, is the Science of Chemistry.

Electricity.—One other agency must here be referred to. Tear up a bit of soft, light paper into very small pieces on the table, and then smartly rubbing a stick of sealing-wax on your sleeve or on any other woollen substance, bring the rubbed end near the bits of paper, and you will see them rising from the table and trying to come to it. There is clearly something present resisting and overcoming their gravitation, and, in fact, we have here the first faint indication of the wonderful power of Electricity.

FRICTION is an agency known only by the resistance to motion. This resistance arises when the rubbing of things against each other tends to check and stop their motion. A ball rolled along the ground soon slackens and is stopped by the rubbing or friction between it and the ground; but, if it is driven along smooth ice, the same energy will carry it much further, because friction is much less on smooth, polished surfaces. Friction, either of solid things, or of water, or of air, must be taken into account in considering movements on our globe, though its whole work consists in resisting and checking other agencies.

Force Defined.—The sense of resistance which our muscles experience when we try to overcome friction, or to set anything in motion, gives us the idea of *Force*. *Any cause which holds things together, or sets things in motion, or alters the speed or the direction of moving things, is called a "Force."*

Perhaps we shall most easily realize the immense and constant work of some of these forces by thinking what the world would be like without them. There is a charming fairy-tale called " The Light Princess," in which that unfortunate young lady is deprived of her gravity, in more senses than one, and therefore habitually floats about the ceilings instead of remaining on the ground, and when out of-doors has to be carefully anchored. If gravity disappeared altogether, not only should we follow her example, but so would our furniture, houses, and everything in the world, and that without the possibility of being anchored, while the world itself would say good-bye to the sun, and depart, cold and dark, into space.

Quite as disastrous would be a state of things without cohesion. If nothing ever held together, but all crumbled down to its smallest particles, the only question would be whether the world would remain as a huge dustheap, or whether it would not rather all fly away in a mixture of gases.

If, however, we had only to dispense with friction, we might indeed still have a world, but a world very like a bad dream, in which we were for ever trying to balance ourselves on surfaces smoother than ice, and to lay hold of things more slippery than eels.

CHAPTER XII.

WE must now try to understand exactly what is meant by the words Work and Energy. *Work means the over-coming through any distance, any kind of resistance.* The greater the distance moved through or the resistance overcome, the greater is the work done. Now the name " Energy " is given to the *power* of overcoming resistance. Hence *energy is the capacity for doing work.* It implies, therefore, not only a *Force* but a *Space*, through which the force is free to act. An illustration will make this clear.

Let us suppose you pick up a heavy ball—say a croquet ball—and carry it up to the top of a church tower. As long as it lay still upon the ground, the force of gravity was not causing any motion ; it was apparently inert ; but, by taking the ball further away from the centre of the earth, we have given the force space through which to act, and now it can produce vigorous motion. At the top of the tower you must hold the ball fast, or it will be off, to a certainty. It now has energy ; but your superior strength prevents it from obeying the downward pull, and so the energy remains stored up in the ball, waiting till it has a chance of acting. Energy stored up like this, waiting for its opportunity, is called *Energy of position* or *Potential Energy.*

If you presently let go and allow the ball to fall, the gravitation will not only *desire* to draw it to the earth but will actually *do* it—as it falls the ball loses the energy of position given to it and gains energy of motion ; it now has *active* or *Kinetic Energy*, as it is called. Any unfortunate person below who gets the ball on his head, or who sees it coming and jumps out of the way, will have a very vivid impression of the active energy which it possesses.

But, perhaps, instead of coming to the ground, the ball may be caught in a tree on the way. What happens then ? The downward pull is not exhausted, but the action is stopped ; the ball still has some energy of position, which will become energy of motion as soon as the ball is set free to move on. When, however, the fall is over, and the ball and earth are together, the gravitation tendency is so far satisfied and is no longer active ; there is no activity left in the ball, its energy is gone somewhere else. But does the gravitation force cease? No, but its effect is different; it does not now produce motion, but it holds the ball and earth together, offering quiet resistance to the attempts of other forces to separate them. It is as if the force between the two, being satisfied, went to sleep ; but it is there, and would be instantly awakened by any attempt to lift the ball.

Therefore things only possess *potential* energy when they have some position of advantage towards something else, and things only possess *kinetic* energy when they have some kind of motion, visible or invisible. If we

want to give fresh energy to the ball after it has fallen
to the ground, we can do so by sending it flying with a
kick, or by lifting it up once more; but in either case we
must give up *just as much of our own energy as the ball
gets.* When you carry the ball up the tower, at every step
of the way you are spending energy, part of which is passing into the ball, and all the energy to fall which it possesses when the top is reached is taken from you. In
ascending the tower your muscles have been at work, and
the muscular energy which you have thus been expending
has necessarily been increased by the weight of the ball
you have carried. Thus the raised or kicked ball derives
its energy from the store you possessed, and this store
was in its turn derived from the food you have digested.

Just in the same way, when water is pumped up into
raised tanks, the muscular or other power that is used
in working the pump gives away to the water just as
much of its own energy as the water will use in flowing
down again; and as long as the water remains in the
tank the energy is stored up in it as potential energy.
Or, if we stop water that is naturally flowing downwards,
and so already possesses energy, by building a dam
across the stream, the gravitation tendency, thus stopped
in the middle of its action, is stored up in the water as
potential energy. In either case the water will certainly
flow down again at the first opportunity, spending its
energy as it does so; and by regulating its opportunities
for flowing, we can settle what the energy shall pass
into as it leaves the water. If a water-wheel is placed in
its path, some of the energy will pass into this, and can
be set to grinding corn or sawing wood; while the water,

when it has reached its lowest possible point, comes to rest—no more work can be had out of it until something energetic comes and raises it up again.

When a coal fire burns and gives out heat, there is a great deal of very active energy at work. Where does it come from ? It comes out of another unsatisfied force, the chemical attraction which exists between oxygen and coal, or rather the substance called carbon in the coal. While coal and oxygen are both cold they have no opportunity of combining ; the force is prevented from acting, and so is stored up in the fuel as potential energy. But when the coal is sufficiently heated by putting burning sticks under it, it begins to combine with the oxygen in the air.

From what we have said it is quite clear that when you turn on a water-tap and set the water running, you do not thereby *give* the energy to the water ; all you have done is to set going the change from the energy of position which the water possessed to energy of motion. In the same way if a ball is caught in a tree, or a stone on the roof of a house, and some one dislodges either, they have simply set free the stored energy which the ball or stone may have had for years undisturbed and undiminished. And so the person who lights a fire does not have to supply the energy used in the burning, because the fuel, being full of stored energy, only needs just so much additional help as will suffice to remove the obstacle which prevented the chemical affinity from acting. Thus the potential energy of the fuel is changed into the active energy of Heat, and in this form it can be set to fresh tasks, or converted again to other forms of energy. If the fire is made under the boiler of a steam engine the heat will set the engine to work; if it is beside the kitchen oven it will cook our food, the energy passing again as it does so into another phase of chemical affinity, and bringing about chemical changes in the food.

We can easily put together long series of transformations of energy. First, potential energy in cold fuel waiting to combine with oxygen, then the same getting its opportunity and changing into active energy as heat under a boiler; the heat gives away its energy to the water, heating it into steam ; the steam works a pumping engine, and at every stroke gives away energy to the pumped-up water, in which it is perhaps stored away again for a time as the potential energy of unsatisfied gravitation. But when the water is allowed to flow and the energy becomes active again, it may be paid away in turning a water-wheel, and if the wheel is employed to drive an electric machine we may presently see the energy reappear in the brilliant form of the electric light.

So energy runs round and round, acting now in this force and now in that, moving things from their places, or altering their shape and condition, making cold things hotter, or melting solid substances into liquids, or combining them chemically into new substances, or producing the activities of light, sound, electricity, etc. Only in the very act of performing each of these works, it is passing away into some other form. For the energy is not destroyed, it is not dead and gone; it does but pass off to other forms of activity, or else is stored up waiting for its next opportunity.

A good illustration of the whole subject is found in its likeness to the use of money. As long as our money is stored up in the bank it is in the condition of potential energy. It is at our command. We have the power to use it as we please, but we are not at the moment using it. When we pay money away for anything, it is gone, so far as we are concerned. We have had our money's worth in the goods which we have bought, just as we had the worth of the energy in the work which it has accomplished. The money is gone from us; we cannot use it again. But it has not perished. The person to whom we paid it has it now, and it may either be stored up again by him, or paid away to a third person for some other goods, and so keep on circulating. And the money may be in different shapes—now in cheques, now gold, now silver, now English money, now foreign, but always representing the same value, and having the same purchasing power.

Fresh money may indeed be coined from time to time, but we cannot ever coin fresh energy; whatever energy is used must be obtained from something that already has it, and so it, like the money, is kept circulating.

An experiment may help us to understand better the constant change from energy of position to energy of motion, or *vice versâ*, which is always going on around us. Tie a small but rather heavy weight to the end of a string, and hang it up so that the string and the weight can swing freely. Then, if you move the weight up into the position of A (p. 213), and there let it go, it will be at once drawn down again by gravitation. The string will not allow it to fall straight down, so that it falls in the only direction it can, back towards B. When it reaches B, the lowest place possible to it, the gravitation moves it no longer; but the weight is full of energy, and as there is nothing to stop its motion, it continues to move in the same direction until it has swung up to C,

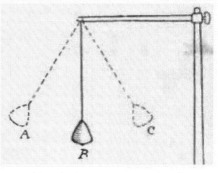

Weight swinging to illustrate alternating types of energy.

as high as it was at A. By this time its energy of motion is used up; but it has gained energy of position, for the weight is now in such a position that it can again acquire active energy from gravitation, so that it is drawn back to B, acquiring on the way momentum which carries it up again towards A, and so the swing goes on.

If there were no resistance it would go on for ever, or, at least, until the string, and that which the string hangs from, were worn out. But as in every swing a little of the energy is taken up in overcoming the resistance of the air, the weight never really rises quite up to the same height again, and so getting lower and lower each time, gradually comes to rest at the bottom point B. When we wind up a clock our muscles supply the energy, which is stored in the coiled spring or raised weights of the clock, and this stored energy is drawn upon to overcome friction and keep the pendulum always swinging, and the wheel-work moving.

CHAPTER XIII.

IF you were standing by a church tower, and some one threw down from the top a cork which fell on your head, you would get a sharp tap from it, but you would not be injured; if, however, a stone of the same size were dropped on your head there would be a very different tale to tell. What makes so much difference between the stone and the cork? We say that the stone is heavier; it has more matter—and therefore more active energy at the end of its fall—than the piece of cork.

Do you remember that when the fallen picture was lifted up, we found that the *weight* of it meant the measure of its resistance to being separated from the ground; that is, of the attraction which the earth has for it? So, if the stone is heavier than the cork, it must mean that it is more strongly attracted to the earth than the cork. Take a large lump of cork in your right hand, and a stone the same size in the left, and hold them up side by side; you will find the stone pressing heavily down towards the ground, and will have to use much more strength to keep the left hand from sinking than the right. The gravitation tendency, then, is stronger in some things than in others; or, to put it more simply, some things are heavier, bulk for bulk, than others.

219

But though the *strength* of attraction, or the weight, may be different, yet the *pace* at which things fall, their *velocity*, as it is called, is the same for all things falling from the same height, unless something gets in the way and resists one fall more than another. This seems a strange thing to say, and we may very likely think that it cannot be true. What? Is it meant that if a stone and a cork and a handkerchief and a feather were all dropped from the top of the tower together they would all reach the ground at the same time? No, certainly not; the stone would come down first; but then we have to consider what they fall through. They pass through the air, and the lighter bodies, in falling, have less energy to overcome the resistance of the air in proportion to their surfaces. It can be proved that any difference in their velocity is due to the surrounding air, for when they are made to fall in a place from which all air is pumped out, then it is found that all bodies,

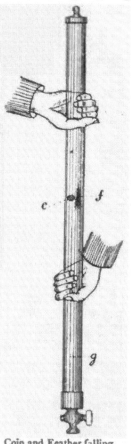

Coin and Feather falling in vacuo.

whatever their weight, will reach the bottom in exactly the same time.

Here is a picture of a glass tube, *g*, about six feet long, closed at one end, and having a stop-cock at the other, and containing a coin, *c*, and a feather, *f,*—any small objects will do. By connecting the tube with an air-pump all the air can be sucked out of it, and if the cock is then closed, and the tube suddenly turned upside down, all the objects contained in it will fall to the other end in the same time. The hands are held in the way shown, so that the tube can be quickly inverted.

But now we come to another point. Instead of travelling all the way down at an even pace, each thing begins to fall comparatively gently, and gets faster and faster all the way down, so that the further it falls the greater becomes its speed. You may see something of this difference in water falling from a height. Get some one to pour water from a kettle out of the first floor window, and look at the falling water. The top of it where it leaves the spout is a gentle continuous stream, but a little below it becomes hurried and uneven in its flow, and presently, as it gets faster and faster, breaks apart into separate drops.

The reason for this quickening of speed is curious and interesting.

We notice in the things round us that not only do bodies at rest remain motionless until some force sets them moving; but bodies once set in motion go on and on until some force stops them. The main stopping force on the surface of the world is friction; but our world itself and the stars, as there is no friction to interfere with them, roll on and on in their paths because nothing stops them. Well, then, when a stone leaves the hand of a person on the top of the tower, the force of gravity starts it moving towards the earth: but if the gravitation force left off acting the instant after, and had no more effect upon the stone, it would still move steadily on because, having once been set going, go it must, straight on in the same direction, until something turns or stops it. Instead of that, however, the force of gravity keeps on urging it afresh every moment of its fall, giving, as it were, fresh and fresh pulls beyond what is needed to bring it down; so all this added force goes to make it travel faster and faster, and the further it falls the greater becomes the velocity or pace of its falling.

It is usual to measure the velocity of a moving body by the number of feet it travels in a second, while the weight of it is measured in pounds. So if we have a ball weighing five pounds moving at the rate of ten feet a second, we should say its weight is five and its velocity is ten; and by multiplying the weight and velocity together we find what is called the *momentum*, which in this case would of course be fifty. An iron ball has more momentum than a wooden ball of similar size dropped from the same height, because its mass is greater.*

Now, when the motion suddenly comes to an end— when the falling body reaches the ground, or the bullet strikes something—what becomes of all its energy? Well, it will do some work or other; but what the particular effect will be depends on the amount of its energy and the condition of the surface against which it strikes. If a heavy body has fallen a great distance, so that its energy is great, it is not unlikely to use its energy

---

* "Mass" is "quantity of matter," and is everywhere a constant property of a body, whereas the *weight* of a body varies with the force of gravitation.

in burying itself in the ground, or in ploughing it up. If it falls into water, the water will be set in violent motion, splashing out and heaving and rippling in all directions. If a hard body falls on a hard unyielding surface the energy will turn to heat and make them both hot; while an elastic body will probably spend its energy in bounding up again.

Density and Volume.—Have you ever seen an old wool mattress, thin and hard with long wear, unpicked to have its stuffing put right again? Out comes the stuffing, squeezed into hard flat lumps and wisps, and then the pickers set to work and pull asunder every lock and curl of the wool till all is light and soft as a snow-drift. But how much room it takes now! As it came out there was not even enough to fill the old mattress bag, and now it has become a mountain of wool several feet high. Is it really the same quantity; and how can we measure quantities which vary so much in size and appearance? Well, one thing has remained the same throughout, and that is its weight; so that by weighing it before and after picking we can make sure that the quantity of matter, or mass, as we call it, is the same.

Our mountain of wool illustrates for us the three things that we must learn to observe in every substance that we have to deal with. First, its mass, which is a constant quantity; secondly, the space which it fills, called the *size*, or *volume*, or *bulk;* and thirdly, the closeness with which the particles are packed together, or the *density*. The quantity of wool being always the same, the volume and the density vary together, but in opposite ways, or *in-versely;* that is, when the wool occupies the largest space, it is most loosely packed, it is least dense; when the packing becomes denser, the size or volume becomes less.

When solid and liquid substances change their density and volume it is generally through being heated or cooled, since heat lessens the cohesion, so that the particles are not held so closely together, making the density less and the volume greater. But we are all well aware in everyday life that, even when equally cool, different substances have very different densities. Look at the grocer weighing a pound of butter. How much larger the pound of butter is than the pound weight in the other scale! That is because butter is less dense than the metal, so that a larger bulk of it goes to make up the same weight; and a pound of isinglass would be much larger again because its density is much less.

Specific Gravity.—It is not enough to say *greater* or *less;* we often require to know how much greater and how much less, and for these comparisons of density it is the custom to weigh things against water, which makes a convenient standard. We have to take an *equal bulk* of the thing to be tested, say a marble, and of pure water and to weigh them both; if we then divide the former weight by the latter, we get the specific weight or specific gravity of the marble, that is the number of times it is heavier than water.

But how shall we be sure what bulk of water is equal to the bulk of what we have to weigh? In the case of liquids this can be found out by first pouring water into a glass vessel up to a fixed mark and weighing it; then after emptying and carefully drying the glass, fill it up to the same height with the other liquid and weigh again. The glass being the same in both cases, the difference in the weights will be caused entirely by the different densities of the liquids. With solids we must use a rather different plan. Suppose we have a small glass ball to test. Let us pour water into our measured glass up to the line marked, say, thirty, and then drop in the glass ball. The water, of course, will rise higher, and if we find that it rises to forty, then it is plain that the bulk of the ball is equal to a bulk of water that fills ten divisions. If the ball weighs 6 ounces and the ten divisions of water weigh 2 ounces, then the specific gravity of the ball is $6 \div 2 = 3$; that is, it is three times heavier than the same bulk of water. This method answers well for small solid things that do not dissolve in water.

For the reason given below, any substance weighed in water appears less heavy than when it is weighed in the air. Let a solid substance, say one of the larger metal weights, be put into one scale of a balance, and weights be added in the other scale until it is exactly balanced; we have then found its weight in the air. Then hanging it on to a hook below the scale, place a jar under it and pour in water until the whole of the weight is under water. We shall find that the balance is no longer even; the water gives more support to the weight than the air did, so that the other scale is now become too heavy. By taking off weights from the other side until the two are again exactly balanced, we can ascertain how much weight has been lost by weighing in water; and we shall find in every case that *anything weighed in water loses just the weight of a quantity of water equal to its own bulk.*

If we now detach the metal weight and let it go, it will, of course, drop down to the bottom of the water, for, though it lost part of its weight, yet it is so much heavier than water that there is plenty left to make it sink. If, however, the experiment were tried with something exactly the same density as water, then, on being detached, it would remain where it was in the middle of the water, neither going up nor down. A substance lighter than water cannot be weighed alone in water, as it will rise up towards the surface and float. If it is half the density of water, it will rise till half of it is above the surface; if only a quarter the weight, three-quarters of it will rise out of the water. To find what such a substance loses in weight when immersed, we must, after weighing it in the air, load it with something heavier whose weight in air and water has been ascertained, and then, weighing them in water, subtract the loss in weight of the heavy body alone from the loss in weight of the two together.

Comparing in these ways the densities of different substances, we find that a certain kind of glass is $2\frac{1}{2}$ times as heavy as water: then $2\frac{1}{2}$ (or 2·5) is called the *specific gravity*, or sometimes the density, of that glass, the weight of pure ice-cold water being taken as unity, *i.e.* 1. Heavy flint glass has a greater specific gravity, and is $3\frac{1}{3}$ times the weight of water. The specific gravity of coal is 1·3, of iron 7·8, of lead 11·4. Among things lighter than water, the specific gravity of wood varies in different kinds from four-fifths to two-fifths the weight of water, while cork is about one-fourth, or 0·25.

As yet we have been speaking only of solid lumps of

substance, the same all the way through, and when this is the case we have only the specific gravity to consider, so that anything with greater specific gravity than water will sink, or with less specific gravity will float.

But now take a hollow ball of thin sheet copper. If all the copper it contains were hammered down into a solid lump it would make but a very small ball, and would be much heavier than a ball of water *the same size*, but when beaten out thin and shaped so as to enclose a large quantity of air, then the ball with the air it contains is lighter than a ball of water *the same size*—lighter, as we say, bulk for bulk, and so it will float in water. This is what makes it possible to build ships of so heavy a material as iron. Were the iron a solid lump it would go to the bottom at once, but the iron shell of the vessel is made to enclose so large a space that it displaces a great deal of water, and hence the vessel, with all its fittings and cargo and crew, still is lighter, bulk for bulk, than water, and so will float.

CHAPTER XIV.

Pressure of Liquids.—If we have a square box containing a block of wood which exactly fills it up, the sides as well as the bottom being everywhere in contact, and then by means of a cover and handle, bring strong pressure down upon the top of it, we shall make the wood press more heavily and closely against the *bottom* of the box, but the pressure will make no difference at all to the sides. The sides might be replaced by the thinnest tissue paper, but it would not bulge or be torn, for the pressure all goes downwards. If, however, the wood were taken out, and the box filled up with water instead, the case would be quite different. Then, any pressure made upon it would press as heavily against the sides as against the bottom, and if the sides were replaced by any thin weak material they would burst asunder : for in liquids pressure is communicated equally in all directions, sideways and upwards, as well as downwards.

To see the upward pressure of water, take any open tube—a common lamp chimney does very well—and close its lower end with a card, which can be held in place for the moment by a thread tied to it and carried

228

up through the tube. Plunge the tube thus closed into a vessel of water, and you may let the thread go, for the card will be firmly held in its place by the pressure upward of the water against it.

When the pressure, instead of being made from outside, is caused by the mere weight of the water itself pressing on its own lower layers, it, of course, increases with the depth of the water. You know that if the tap of the water-butt is opened when the butt is full, the pressure of all the water above not only forces water out of the tap, but gives it so much momentum that the stream spurts out in a horizontal direction for a short distance before gravity draws it down to the ground (as at A in the picture).

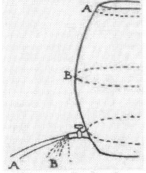

Water spouting from butt. But when the butt is only half full, the water has less momentum from pressure, and reaches the ground in a much shorter curve, B; while, when it is nearly empty, there is a mere trickle from the tap, with no spurting power left in it.

We have seen that the top of any liquid, when at rest, is always a level surface, and this is true not only when it is all contained in one vessel, but in any number of vessels of any shape or size, provided they all communicate with each other below. Thus the water always stands at the same level in a watering-pot and its spout. The watering-pot is like a large tube, and the spout, a small one, opening into each other below; and you cannot raise the height of the water in one without raising it in the other also. Of course, if one of the tubes is lowered until it is below the level of the water in the other the water will run out until both stand again at the same level. This is almost too obvious to need saying, since every cup of tea is filled by the simple process of lowering the spout of the teapot until it is below the level of the tea in the pot. But we may not have noticed the same thing happening on a larger scale. A tank of water at the top of the house, with a pipe leading down from it to the garden, and there turned upwards again, form together just such a vessel as the

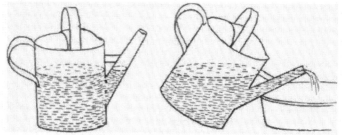

Uniform level of water in can and spout.

teapot with its spout lowered; and, in this case, not only will the water overflow at the end of the pipe, but the pressure of the water above will give it so much momentum that it will spring up in a fountain.

Pressure of Gases.—Air has weight. If it were not acted on at all by gravitation, then, since the molecules of gases are always repelling each other, the air would presently fly away from the earth altogether, and go off into space; happily, however, its attraction to the earth is strong enough to keep it here. Indeed, the sea of air above us is so deep that its weight causes a very considerable pressure upon everything on the surface of the earth, amounting on an average to about fifteen pounds on every square inch, and so to several tons on the body of a man.

Why, then, do we not feel weighed down by this great pressure? If air pressed like a solid substance, downwards only, we should be crushed under it, but as, like a liquid, it presses equally in all directions, the sideway and upward pressures balance the downward pressure, and, among the balanced forces, we move as freely as a fish moves through all the balanced pressures in water.

When, however, anything causes a slight difference in pressure in different directions then we do feel the pressure of the air. If it is very slight we say there is a breeze, but if the inequality becomes greater the pressure may increase up to a terrific gale, before which great trees go down like nine-pins.

Some gases are heavier than others. Hydrogen gas, for instance, is a good deal lighter than the mixture of gases that we call air, and consequently it will rise and float in the air-sea, as cork rises and floats in water; and just as a copper ball, when enclosing a sufficient quantity of air, can float in water, so other materials may be arranged to enclose hydrogen enough to make the whole mass lighter than *an equal bulk* of the lower layers of air, when it will rise and float in the atmosphere. Such a ship of the air is called a balloon.

We employ the pressure of the air to do useful work for us in various ways. Look at this picture of the inside of a pump. You see the barrel or body of the pump, with a tube below which dips down into the water, and a thing inside, called the piston, A, which

moves up and down as we move the pump handle. The
piston fits quite closely into the barrel of the pump so
that no air can pass up or down round it, but it has in
the middle a little door or valve which can only open
upwards. Another valve B,
also opening upwards, closes
the entrance of the water pipe
into the bottom of the pump
barrel. Suppose now that the
piston, which is connected with
a rod and handle, is drawn up
to the top of the pump, and by
moving the handle we begin to
press it downwards. What
happens? The space below
it is full of air, and when the
piston coming down presses

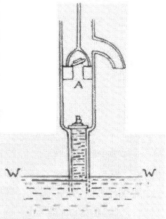

Diagram of Suction-pump.

upon this air the enclosed compressed air opens the
upper valve, shuts the lower, and so escapes upwards.
When the piston is at the bottom there is not any upward
push of the air and the upper valve falls shut, and as
the piston is drawn up again, the air above pressing on
the top of the valve keeps it firmly closed so that none
can return through it. The inside of the barrel is now
partly emptied of air, but it does not stay empty, for
first the air, and then the water in the pipe, rushes in,
pushing open the lower valve B, and following the rising
piston. Why does the water rush up into the tube as
the piston rises? The surface of the water, WW, *outside*
the tube is bearing the pressure due to the atmosphere
above it, and as long as the tube was also full of air the

pressure was balanced, and the water remained level. As soon, however, as some of the air is withdrawn from the tube, the pressure outside is not balanced and presses the water into the tube to occupy the space left vacant. Another stroke or two of the handle pumps out all the air so that the water rises up to the piston, and then when the piston descends again it is water and not air that is forced through the upper valve; lastly, on raising the piston the valve closes, the water is lifted and flows out of the spout.

We can see that if the barrel of the pump had no tube below to go into the water, but was all solidly closed with materials strong enough to bear the pressure of the air without breaking, we could then, by pumping out all the air, have a really empty place inside, a vacuum, as it is called.

The water-pump will work perfectly well as long as the tube is not too long, but if the distance of the piston above the water is more than about thirty-three feet it will fail, for the water will not rise so high; because a column of water thirty-three feet high has become heavy enough to balance the pressure of the atmosphere, and when the forces are balanced no more work can be done. It is therefore plain that a column of air, say a square inch across, the whole height of the atmosphere, has just the same weight as a similar column of water thirty-three feet high. This is how we know that the atmospheric weight or pressure is fifteen pounds on each square inch of surface, for this is the weight of a column of water a square inch across and thirty-three feet in height.

With a heavier liquid than water it would, of course,

take a shorter column to balance the weight of the atmosphere, and mercury, the heaviest liquid known, balances it with a column of about thirty inches. We say *about* thirty inches because there are variations in the pressure of the air; if the pressure of air is somewhat greater, it can balance a rather higher column of

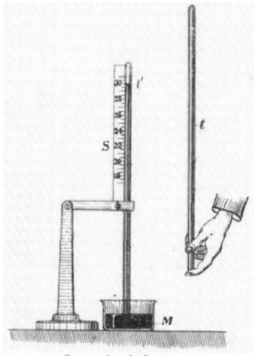

Construction of a Barometer.

mercury, or, if it is less, the mercury balanced by it will fall a little. This is the explanation of the barometer.*

* Barometer—from two Greek words meaning "measure of weight." As mercury is 13½ times denser than water, the height of the mercurial barometer is 13½ times less than a water-barometer; 33 feet or 396 inches ÷ 13½ = 30 inches nearly.

A barometer consists of a glass tube more than thirty inches long, closed at one end. To prepare it, mercury is first poured into the tube till it is quite full, and then the operator, stopping the open end with his finger, as shown in this picture at *t*, turns the tube careful'y upside down and dips the end into a vessel of mercury, M. When he withdraws his finger, the mercury falls a little way, as at *t'*, because the column is more than thirty inches high, but it stops as soon as its weight is balanced by the pressure of the air on the mercury in the open vessel, and as the empty space left above it in the top of the tube is a vacuum, the column rises and falls with every variation in the weight of the air, these changes of height being measured upon a graduated scale, S. When we say that "the glass is high," or "the glass is falling" (meaning, of course, the mercury in the glass), this shows that the pressure of the air is greater in the former than in the latter case, and as rain rarely falls when the atmospheric pressure is great, the barometer is given the name of the weather-glass.

CHAPTER XV.

IF we take a good lump of ice, and put it into a saucepan in front of the fire, the outside of it will first grow moist and melt, and drip into the bottom of the pan, then the next layer will go in the same way, until the whole ice is slowly changed into water. The ice stood up in a lump of uneven shape, reaching above the top of the pan but not filling the bottom. The water, on the contrary, runs down and fills all the bottom and sides till it is exactly of the shape of the pan, only with a flat, level, upper surface.

Now, let us put our saucepan of melted ice on the fire. The water gradually becomes hotter and hotter, and steam begins to rise from it into the air. Presently it bubbles, and heaves, and boils, and throws off steam in clouds; do not take it off the fire, but watch what happens. It continues to boil, but there is less and less of it in the pan, and if allowed to remain over the fire every drop will disappear, and the pan be quite empty and dry. What becomes of the water? Where is it gone? It is all changed into vapour, and as such gone away into the air.

If instead of boiling it in an open saucepan, we put it

into a kettle with a tightly closed lid, the steam and vapour will pour out of the spout. Let us lengthen the spout by fixing to the end of it a good long tube, as in the picture, and wrap round the tube a cloth wrung out of cold water. This cools the vapour on its way through the tube, and as it loses its heat it is condensed, that is, it changes back again into water, which drips from the end of the tube into the basin set underneath. We shall have to keep on cooling the wet cloth as it gets warm, and gradually the basin will fill with water, which came out of the kettle in the condition of vapour. Supposing

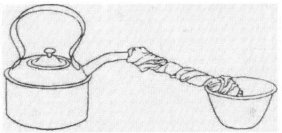

Water distilled from a kettle.

this experiment to be made during a sharp winter frost, it may be completed by setting the basin out-of-doors where the water will gradually grow colder and colder until it is frozen into ice again, as at first.

Now, in this experiment, the *substance* of the water is the same all the time, but we see it pass through three different states or conditions. First, it was in a solid state of ice, then, by heating, it was changed into a liquid state of water, and heating still further turned it into a state of vapour or gas. And afterwards some of it was cooled down again from vapour to water, and from water back to ice.

When we come to think of it, all the substances that we know in the world exist in one or other of these three states—solid, liquid, or gas. We could name plenty of solid substances, such as iron, and wood, and wax; there are liquids, such as water, and oil, and quicksilver; and there are gases, such as the coal-gas we burn, and the air we breathe. The water was changed from one condition to another by making it hotter or colder, and water is a very convenient example to use, because we are all so familiar with it. Every winter, as a rule, in this climate, there is cold enough to freeze it into ice and snow, and any fire gives heat enough to turn it into vapour. But changes of the same kind, though generally requiring more heat, are constantly made in other substances. Lead and tin are melted, or fused, as we say, without much difficulty, and also no one who has seen an iron foundry can ever forget the look of flowing streams of molten iron, quivering with heat. Although there is a very great difference in the readiness with which different substances will change their condition, yet so many have actually been changed by experiment, that it seems reasonable to believe that all might be made to pass into the other states if we could get heat or cold enough to act upon them. Even air has been made liquid and some gases solid by great pressure and intense cold.

But let us understand exactly what we mean in saying that probably all solid substances might be melted or vaporized, and all liquids turned into vapours, if they were made hot enough. There are plenty of solid things all round us. Here, for instance, is a lump of coke; could

I melt that by heating it? Coke consists chiefly of carbon, which is one of the most difficult substances in the world to fuse. But long before it became hot enough to melt, something else would happen. A very moderate degree of heat would enable the carbon to combine with the oxygen in the air, and it would all burn away, and pass into the state of gas by chemical combination. It would be necessary to prevent any oxygen getting to the hot carbon if we were really to try and melt it; so that we should not succeed by merely using great heat without thinking of other circumstances.

Cohesion.—The real difference between the three states of matter is a difference of, or absence of, Cohesion. The cohesion is strongest in solid bodies, and the most perfect solids are those like iron or granite, in which cohesion offers the greatest resistance to the other forces which would change their shape or size. Take the poker from the fireplace and examine it. How strong and solid it is! It will neither break, nor bend, nor stretch, nor squeeze with any force of our hands. Many solids, however, can be readily broken or bent, and, in fact, we can find very different degrees of firmness among them, but we call every substance solid which has cohesion enough to keep some shape of its own, resisting the force of gravitation which tries to draw its upper particles down among the lower. In this sense even a lump of jelly is a solid, yielding as it is in other ways. Some substances, like pitch and cobbler's wax, that look solid, and are even brittle to a blow, gradually yield to the force of gravitation, and flow like a thick liquid if time enough be given : such bodies are called viscous solids.

In liquids the cohesion is very slight, the molecules moving readily over each other in any direction, so that they are free to follow the guidance of gravitation. When we first put the lump of ice into our saucepan, you remember that it stood up with an irregular shape of its own; but when it melted and became liquid, gravitation drew the obedient drops trickling downwards into the bottom of the pan, and allowed none to stand up above the rest, so that, when all was melted, the top of the water presented a level surface.

In gases or vapours (vapours only mean gases easily turned into liquids) there is no cohesion at all. So far are their molecules from holding together, that they are always trying to spread themselves as widely apart as possible, and thus gases readily fly off, and are diffused into the air. In fact, if they are to be kept separate at all, they have to be carefully bottled up.

So then, if the cohesion in a solid substance is sufficiently lessened, the solidity disappears. If a little cohesion is left, we shall have a liquid, or, if the cohesion is quite destroyed, the substance will fly off as a gas.

Heat, as we know, is the great enemy of cohesion, and no solid or liquid can hold together if there is strong enough heat separating its particles; but it does not follow that by taking away the heat we can always bring back the cohesion. The molecules of vapour may get too far apart for cohesion to act. The vapour that went off from the boiling saucepan was lost in the air, and no one could bring it together again to condense it back into a draught of water. It cools down, of course, in the cool air, and, though some may be condensed again, in the form of rain or dew, yet much may remain in the state of diffused vapour even when the air is cold enough to freeze water.

Evaporation.—Wherever water, or even ice, has an uncovered surface (either in air or a vacuum), some of it is quietly and invisibly passing off in the form of vapour, unless the air is so full of water-vapour already, so damp, as we say, that it can hold no more. We do not notice this much when there is a large body of water, but when we hang up wet clothes in the air to dry we reckon that the water in them will soon *evaporate*, or pass off in vapour. If the air is very damp, so that it will not take much more water-vapour, we say it is a bad drying day. Hot air can take more vapour than cold air, so that in hot weather, or in front of a fire, things will dry quicker; but, then, if anything cools the air it can no longer hold so much, and some of the vapour condenses into water again. So, after a hot, drying day in summer, when the sun sets and the air cools down, we are apt to have a heavy dew, which is just the extra vapour in the air deposited in the form of water upon the cooled grass. But next morning, when the sun has been up some time and the air is warmed, it will drink up the dew again.

Solution.—Here is a solid lump of sugar. Put it into this cup of water, and after a short time it will have disappeared. The cohesion that kept it solid is lessened by the affinity that water has for the sugar, and it has passed into the form of clear liquid, which mixes with the water. The sugar is *dissolved*, and the mixture is called a *solution* of sugar.

241

Now, if we go on putting more and more sugar into the cup, it will continue to dissolve for some time, but at last the water will have taken up as much as it can hold; and if more sugar is added after this point is reached it will not be dissolved at all, but will remain solid. The solution is then said to be *saturated*. If a saturated solution is kept in an open vessel where air can reach it, the evaporation which constantly goes on at the surface will gradually draw away the water, and the dissolved substance, now become too much for the diminishing quantity of water, will slowly reappear in a solid form.

Crystallization.—And here, for the first time, the force of cohesion, or molecular attraction, is seen doing active work. Hitherto we have only noticed it quietly resisting the active work of other forces ; but here, as the molecules are gradually set free from the water, each begins to be drawn into its place and built into the solid mass. And how wonderful and beautiful is the building ! When the attracting force is not interfered with in any way,—that is, when the solidifying is slow enough, and there is plenty of time, and plenty of room for the work, and the solution is quite still and unshaken,—the new solid will frequently appear in the form of regular crystals, often clear and transparent, small at first, but gradually increasing in size as fresh layers of molecules are deposited ; and every crystal has its own fixed form, always the same for the same substance.

Take a lump of alum, and put it into a tumbler with just cold water enough to cover it. In a few days we shall find its surface eaten out into a variety of forms more or less regular. If we now put a drop of the water on to a slip of glass, and set it on the chimney-piece in very gentle warmth, we shall find, as it slowly evaporates, that the alum dissolved in it reappears in the form of tiny, eight-sided crystals, which may be

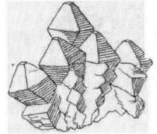

easily examined with a magnifying glass. Some of them will very likely look as if their corners or their edges had been cut away, but all preserve the same general form; and we recognize the same forms also carved out by the water in the

Group of alum crystals.

larger mass of alum. Here is a picture of a big group of alum crystals, as they are formed in the slow cooling of a strong solution.

Dissolve a teaspoonful of common salt, of Epsom salt, and of nitre, each in a different wine-glass of water, and let them stand till next day. Then put a drop of each as before, on slips of glass, where they may evaporate slowly. As they dry up the salt will appear in tiny

A crystal of alum.   Common Salt.   Epsom Salt.   Nitre.

cubes, the Epsom salt in four-sided prisms, and the nitre in six-sided prisms. Here we can compare crystals of different forms, each form, however, belonging invariably to its own proper substance.

Look at a lump of loaf sugar and a piece of sugar

candy. They are both sugar; but why do they look so different? The candy has crystals sharply and regularly formed, while the lump seems to be a mere mass of crystalline grains crowded together, in which we cannot make out any regular forms. This difference depends on the rapidity or slowness of the evaporation. If the solution is rapidly and hurriedly evaporated, the crystals produced are confused and minute; to get them perfect in form, the solution should be kept very still and comparatively cool, that it may evaporate slowly.

The following pretty experiment is sure to please those who carry it out carefully and patiently. Take some fine wire with a little thread wound round it, and twist it into the outlined form of a small basket. Then make a saturated solution of alum in water in the following way. Take rather more boiling water than will be needed to cover the basket, and dissolve powdered alum in it until it will dissolve no more. Strain the solution through a piece of muslin, when it should be poured into a wide-mouthed jar. Tie a loop of thread round the basket handle, pass a piece of stick through the thread and gently lower the basket into the solution. The basket must be completely covered by the liquid, but must not touch the sides or bottom of the jar; it will be kept in its place by resting the stick across the top of the jar. Then put the whole away for four or five days into some place where it will not be touched or shaken; if possible it should be locked into a closet where there will be no sweeping or dusting. The crystals of alum will be deposited in clusters on the wires, turning them into a crystal basket, but the whole being evenly covered depends chiefly on the jar remaining entirely undisturbed during the process. If a little colouring matter is mixed with the solution, we can obtain coloured crystals

CHAPTER XVI.

HEAT.

Expansion.—We have already had occasion to talk a little about Heat while studying Cohesion and Gravitation, and have seen that it is often at work overcoming Cohesion, and making the density of substances less and their size larger. Here is an experiment to show us a solid substance growing larger, or expanding, with heat.

Take a common steel knitting-needle, K, K (p. 242), and with a vice or clamp, *c*, fasten it in a horizontal position and, securely at one end, to a block of wood, *b*. Get another block of wood or a book, *b'*, and on it place a bit of window glass, *g*, and on the glass a sewing-needle, *n*, the end of this needle is only seen in the picture. Fix a straw, or any light index, *i*, on to the point of the sewing needle, and hang a letter-weight, W, to the free end of the knitting-needle to keep it pressed down. If the knitting-needle is now heated by putting a spirit-lamp, L, under it, it will expand and grow longer; but the change of length would be too small for us to see, were it not that the end of the needle in advancing rolls the sewing-needle round, and so makes the long index swing round quite perceptibly. It is a good plan to set up behind the block a screen with a straight upright line

246

drawn on it parallel with the index, *i*, so that we can measure how much the index moves away from the upright.

In this case we have a very small steel rod lengthened by heat, and the large iron rails of the railways lengthen in just the same way in hot weather. If their ends were fixed close to each other while they were cool, then their expansion, when it came, would make them thrust each

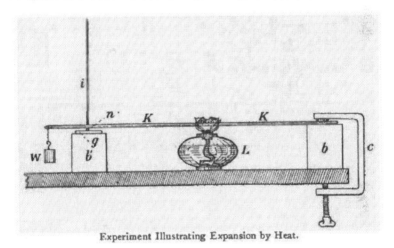

Experiment Illustrating Expansion by Heat.

other upwards or sideways, and so break the straight line of the rails. But next time you are near a railway, go and look attentively at the rails where they join, and you will find that there is always a space left between them to allow for their lengthening.

To see a liquid expand with heat, let us examine an ordinary THERMOMETER.* You see that at its lower end

* Thermometer—"measure of heat," though, as we shall see directly, the thermometer is really a measure of *temperature*, not of the *quantity of heat* in a substance.

is a hollow glass ball or bulb, with a narrow glass tube
rising out of it, and that the bulb and part of the tube
contain mercury. As in the barometer so here the
closed end of the tube above the mercury has been
emptied of air. The bulb and tube are usually set in
a frame, on the back of which is a scale of figures. Hold
the bulb in your warm hand, and as the warmth reaches
the mercury it expands and rises higher in the tube;
put it in front of the fire and it will rise still higher; you
can note the height on the scale at the back. Plunge
it into cold water; the water withdraws heat from it, and
the mercury contracts and falls.

In order to make a scale for measuring the amount
of rise and fall, the thermometer is plunged first into
melting ice for a few minutes; and when the mercury
has left off falling, and is quite steady, its height is clearly
marked on the frame. Next, it must go into boiling
water—not merely very hot water, but water actually
boiling and bubbling. Up rushes the mercury at first,
but presently it comes to a stand and will rise no higher.
When it is steady, this height also is marked. These two
fixed points are called the freezing point and the boiling
point.*

But now comes the practical difficulty in our use of

* As the boiling point varies slightly with the nature of the vessel
in which the water is boiled, and considerably with the elevation
above the sea level, the *true* boiling point is that given by the
steam from water boiling in an open vessel at the level of the sea,
when the barometer is standing at 30 inches high. For as the
pressure of the atmosphere decreases the boiling point falls. On
the top of Mont Blanc, for instance, which is nearly 16,000 feet
above the sea, water boils at 185° F., whereas at the sea level it
boils at 212° F.

thermometers, which is that instead of all agreeing to use one scale of figures, people use two or three different scales. One plan, which seems the most sensible, is to call the freezing point o, and the boiling point 100, and to divide the space between them into 100 equal parts or degrees. A thermometer marked on this plan is called a Centigrade * thermometer, and is generally used in scientific work. But unfortunately on the thermometers in common use in this country the freezing point is marked 32, and the boiling point 212, the space between them being divided into 180 degrees. This is called the Fahrenheit thermometer from the name of the inventor, and you can see that nine degrees of the Fahrenheit thermometer are equal to five degrees of the Centigrade. It is necessary to be acquainted with both of these scales, and we generally find that measurements of heat in books are marked C. or F., to show which is intended to be used. Thus $100°$ C. = $212°$ F. must be understood to mean, one hundred degrees Centigrade represents the same temperature as two hundred and twelve degrees Fahrenheit. If the warmth of your sitting-room is $60°$ F. it will mark not quite $16°$ on a Centigrade thermometer. Notice that the word degree is indicated by the little sign ° placed after the figure, thus $16°$ C., means 16 degrees Centigrade.

Temperature.—Now that we have our thermometer, let us see exactly what it is that is measured by it. It is not the amount of heat that any substance has received, but what we call its *temperature*, or the degree of hotness which it has reached. We will try to

* Centigrade, hundred steps.

249

understand this quite clearly, as mistakes are often made about it.

If we put side by side on a table a small medicine bottle, a tumbler, and a glass dish or pan, and pour into each of them exactly the same quantity of water—say, a small teacupful—it will fill up the bottle to the top, the tumbler will be half full, while the dish will barely have the bottom covered (see picture). The *quantity* of water is the same in each, but the different shapes of the vessels, and especially the different size of the bottom,

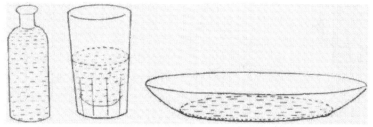

Experiment illustrating analogy of water level and temperature.

makes all the difference in the level to which the water rises.

If we next place in front of the vessels a screen with a narrow slit in it, which will only allow a narrow upright section to be seen, and then invite some one who is not in the secret to look at each vessel through the slit, he will certainly think that the bottle contains the most water; and if we assure him that the same quantity was poured into all, he may say, " Well, I do not know where the rest of the water is gone, but it certainly is not filling them up to the same level."

Now this is exactly as much as the thermometer can tell us about the heat of substances. It cannot tell how much heat they may have received, or what else it may be doing, but can only show the *temperature*, or the level at which the intensity of the heat in them stands.

Again, if we were to stand the bottle of water in the dish, and then knock a hole in the side of the bottle (near the bottom), the water would flow out until all in the bottle and dish stood at the same level. Just so, when we bring into contact two bodies of different temperatures, the heat will pass from the higher temperature to the lower until both stand at the same level.

Let us look at some cases where the same amount of heat applied raises temperature to different levels. Put on to the same fire, and as nearly as possible in equally good positions over the hot coals, two saucepans of the same size, one quite full of water fresh from the pump, and the other with merely an inch of the same fresh water in the bottom, and place a thermometer in each. See how quickly the temperature rises in the inch of water; it will reach the boiling point before the other is warmed through, and when it does so we will take both of them off the fire. They are now exactly in the same position with regard to heat as the bottle and dish with regard to the water. Both have received the same amount of heat, but one is raised to a high level of temperature, while the other, owing to the larger quantity of water over which the heat had to be spread, stands at a much lower level.

Specific Heat.—In this case the difference of temperature depends only on the different quantity of liquid to be heated; but we will now vary the experiment thus. Take two small vessels of equal size and thickness (the small glass fish-bowls sometimes used for flowers are very convenient), and put into one a certain weight of water, and into the other the same weight of turpentine. Place a thermometer in each, and make sure that they are both at exactly the same temperature. If they are not so at first, let them stand near together in the same room for some time, and they will both have the temperature of the air in the room, for whichever is hotter, air, water, or turpentine, will give up some of its heat to the others until all stand at the same level. When they are equal, put them both together, with the thermometers still in them, into a pan of warm water. You will find that when the thermometer in the water-bowl has risen one degree, the thermometer in the turpentine will have risen about two degrees or a little more.

Here the quantities by weight of water and turpentine are the same, and the quantity of heat given to each is the same, yet it takes more than double the amount of heat to raise the temperature of water one degree than it does to raise the temperature of turpentine one degree.

If we tried other substances, we should find that they vary a good deal from one another in the readiness with which their temperatures can be raised; but they may all be measured against water, and the heat that each requires to warm it, say one degree compared with the heat an equal weight of water requires, is called the *specific heat* of the substance—just as the weight of each, compared with the weight of an equal bulk of water, is called the specific gravity. No other solid or liquid substance requires so much heat to raise its temperature as water does, so if we call the specific heat of water 1, we shall find the specific heats of all these other sub-stances to be less than 1, so that they must be expressed by fractions. The specific heat even of ice is less than half that of water, or 0·5 ; of turpentine is 0·43 ; of iron 0·11 ; of mercury only 0·033, or one-thirtieth of an equal weight of water, so that mercury is an excellent indicator for the thermometer, as its temperature rises so rapidly with small additions of heat. We must remember, how-ever, that mercury is 13½ times heavier than water. It is on account of its regular expansion through a wide range of temperature—owing to its boiling at a very high temperature, and freezing at a low one—that mercury is usually chosen for thermometers.

Fusion.—LATENT HEAT.—We have already tried ex-periments in turning ice into water, and water into vapour (see p. 231) ; but, now that we understand the meaning and use of the thermometer, we will take them over again, making more exact observations. Before bringing in the lump of ice let us make a hole in it to hold the bottom of our thermometer and note the temperature : usually it will be 32° F., the freezing-point of water, unless the air is very cold and below the freezing-point. Now put the ice in a pan near the fire and watch the thermometer, it remains stationary. The heat continually added to the ice is no longer raising the level of its temperature ; it is doing other work, breaking down the crystalline order and overcoming the cohesion—in other words, it is melting the ice ; and until every particle of the ice is changed into water, we shall see no alteration in the thermometer.* There are, however, other ways of

* The ice and water must be kept well stirred through this time.

ascertaining the amount of heat which is taken in all through the time of melting; and it is known that the ice takes as much heat to melt it as would serve to raise the temperature of an equal weight of water nearly 80° C. or 144° F., that is, from 32° to 176° F. When all this heat has been expended in overcoming the internal forces, and the whole ice is converted into water, the thermometer in it still stands at 32° F. But the instant this work is done, any further heat added now affects the level of temperature, and the thermometer again begins to rise.

We will now put the saucepan of water on the fire and the rise steadily continues until the thermometer indicates 212° F. (100° C.) or boiling-point. Here, again, it stands still, and never rises above this point while the whole of the water is being converted into vapour. The additional heat now taken in for this purpose is enormous. To turn one pound of boiling water entirely into steam at the same temperature requires as much heat as would raise 965 pounds, or 8½ hundred-weights, of water 1° F. This amount of heat would raise nearly 5½ pounds of water from the freezing-point to the boiling-point. But none of the heat used in changing the pound of water into steam becomes apparent in the temperature; it is at work destroying cohesion, and setting the molecules absolutely free from each other's attraction. You know that evaporation from the surface was going on before, and you have seen steam rising from the water all the time it was heating; but the difference when the boiling point is reached is that the temperature rises no more until the whole liquid is converted into vapour.

Heat which thus passes altogether out of sight, unrecognized by the thermometer, is given the name of *latent,* or hidden, heat.

But when the process is reversed—when vapour is cooled and condensed into water, it sets free again all the heat, the latent heat of evaporation, that was taken up in vaporizing it, and when water freezes into ice it sets free the latent heat of fusion which was before absorbed in melting it, and the heat thus set free becomes again what is called sensible heat, that is, heat that can be felt.

This makes the freezing take place very gradually. When water is chilled down to a temperature of $32°$ F. it does not all suddenly become solid; but as it begins to solidify heat is set free in the proportion of $144°$ F. for every pound of the water, and the whole of this must be taken away by the chilling process before the work of freezing is finished. On the other hand, when ice is melting, it does not all suddenly become liquid when the thermometer rises above freezing-point; but before the melting is finished it must find for every pound of the ice as much heat as would raise the temperature of a pound of water $144°$ F. And fortunate it is for us that the work has to proceed so slowly, else every slight frost would deprive us of all our water, and every thaw after frost and snow would mean a devastating flood, with no time for the water to run off as it melted.

Density of Ice and Water.—Solid bodies, as they liquefy under heat and have their cohesion made less, generally become less dense at the same time, the molecules moving farther and farther apart and being

255

spread over more space, so that the liquid form of a substance is almost always lighter than an equal bulk of the solid form. But two or three substances, of which water is the most easily observed, are densest when they are at a temperature rather *above* their freezing-point, so that ice is lighter than an equal bulk of ice-cold water. The largest number of men may be squeezed into the smallest space when they are crowded together anyhow ; but if they are ranged in regular order like companies of soldiers they will take more space, or if the crowd move apart a little, it will be less dense and also take more space. Let us think of our water molecules as the men. They are densest and most closely crowded together at a temperature of about $39\frac{1}{2}°$ F.; but as the temperature falls to freezing-point they are ranged into their crystalline order, and so expand and take more room, or as the temperature rises they move apart and also take more room, and so the density becomes less either way.

It is a happy thing for the world that ice is lighter than water. Only think what would happen if it were not so ! As soon as the surface of water became chilled by the cold air and froze, the ice, if it were heavier, would sink to the bottom and leave the next layer of water on the top to be frozen and sink in the same way, till the whole mass of water was changed into solid ice in which no fish could live. As it is, however, the layer of ice remains floating on the top and makes a shield for the water underneath from the cold air.

Force of Expansion.—But ice is lighter because larger than the unfrozen water, and when it wants more room it will have more room. If water in our water-pipes is chilled enough to expand into ice and wants more room for the crystalline arrangement, it will make nothing of bursting open the iron pipe to obtain it; and yet the difference of size between ice and water is not so very great. What, then, must be the expansive force of water changing into steam, for their difference of size is enormous! When the molecules are quite set free from cohesion, the water expands into vapour 1700 times its own volume; a pint of water will make 1700 pints of water vapour. Have not you noticed how, when water is boiling in a covered saucepan, the steam keeps lifting the lid and letting itself out in clouds far larger than the water it came from? Every kettle-lid has a hole in it to let out steam; but if the steam is formed rapidly it needs more room and keeps the kettle-lid dancing in the same manner. But what would happen if the lid would not open and the hole and spout were stopped? There would soon be an end of the kettle; and the expansion of steam, not content with merely cracking it open, as the ice did to the water-pipe, would explode it in all directions with such force as to wreck everything round it.

We are not, however, to suppose that it is *impossible* to prevent these expansions; since both ice and steam are hindered from forming by sufficiently strong pressure; but the strength of the expansive force can be estimated by the great strength of materials that is required to resist it, and boiler explosions take place often enough to remind us what a giant we have to deal with in imprisoned steam.

Convection.—How does heat spread through water?

We put cold and hot water together to fill a warm bath, and bring them to the same warmth throughout by rapidly stirring them together and mixing them with the hand. Would they have mixed themselves if left alone? Yes, if the hot water were at the bottom and the cold at the top, for the hot water is lighter and would rise and float while the colder water sank, and as the currents pass and touch one another, heat would be given up from the hotter water to the cooler until both were equal. But no, if the hot water were on the top to begin with; for then there would be nothing to set any currents moving in the water.

Put a saucepan of cold water on the fire with a little soaked sawdust in it to show the currents, and as the bottom water gets warm first, you will see that it is always rising up, while the colder water above is ever sinking down to take its place and be warmed in its turn.

If we have a cup of tea too hot to drink, and simply let it stand to get cool it will cool from the top, the cooler tea will sink while the hotter rises, a current will be set up mixing it thoroughly, and the tea will be uniformly cooled through. But if we are in a hurry, and stand the cup of tea in a deep saucer of cold water to chill it, we have carefully to keep stirring it with a spoon, for now the bottom tea being cooled first, it will lie still at the bottom, and no current will be set up unless it is stirred.

This mixing of hotter and colder parts by the movement of currents is called *convection*, and is the usual way in which heat spreads through liquids and gases.

Every cook knows that she must not leave a pan or kettle on the fire without some water in it. As long as it contains water the heat which the fire pours into it all passes into the water, first raising its temperature, by means of convection, to boiling-point, and then turning it into vapour; but as soon as the water is all vaporized and gone, the heat will set to work upon the kettle itself, and soon make a hole in it. Convection currents carry off the heat so quickly that water may even be boiled in paper. This pretty experiment is easily carried out. Make a square sheet of writing-paper into a deep tray by turning up the edges all round and folding over the corners, put a little water in it, set it over a spirit-lamp with a small wick so that the flame touches only the centre of the paper, and the heat will boil the water without burning the paper. A stand to carry the tray may be readily contrived by supporting a wire ring on three legs of twisted wire.

Conduction.—Go into a room without a fire, and lay your hand by turns on the carpet, the table, the stone chimneypiece, and the iron fender. Which feels the coldest? The carpet does not feel cold at all, the wood hardly at all, while the stone is decidedly cold, and the iron colder still. But they have all been standing there long enough to have the same temperature as the air of the room. Let us fetch a thermometer and test their temperature. It stands at the same level in all. Is there, then, anything in the room of a different temperature? Yes, you are hotter, and your hand is hotter than the rest of the things. When you lay it on the iron the nearest particles are warmed by your warm hand. How does the heat spread through the iron? It cannot be by moving currents, for in solid bodies each particle keeps its own place, and they cannot be driven about among each other like streams of water. In this case the particles that are warmed first pass on some of their heat to the next, and these again to the particles beyond them, thus gradually spreading the heat through the whole mass by what is called the process of *conduction.*

Iron particles pass on the heat rapidly, taking up more and more, and trying quickly to bring your hand and the iron to the same low level of temperature. The stone tries to do the same, but is not quite so active in the work, so that it does not chill you down so fast. The wood is decidedly slower about it, and when the particles nearest the hand have reached its temperature they are content to stay so, without hurrying to pass on the heat. And the particles in the woollen carpet hardly move in the matter at all.

So we see that just as substances differ in the amount of heat required to raise their temperature, so they also differ very much in the rate at which they pass on their heat. Those which pass it on rapidly are called good conductors. But some are so slow about passing on heat, that one part of the substance may be actually burned before the rest of it is heated at all. These are bad conductors. When we want to light a candle, we either strike a match, or twist up a paper spill and light it at the fire ; but, in either case, not only is one end of the match cool enough to hold in the hand while the other is burning, but we can still hold it until the flame almost reaches the fingers. It is plain that wood and paper are bad conductors.

Wool is one of the worst of conductors, and therefore it makes the warmest clothing; for while it does not itself conduct away the heat of our warm bodies, it makes a thick screen preventing the cold air from approaching to rob us of heat. But, for the same reason, it is the best protection for anything cold, and we wrap ice in flannel to prevent its melting, for the flannel does not readily allow anything else warm to come near, from which the ice could borrow heat to melt with.

The best conductors are the metals, but they are not all equally good. You know that iron is a pretty good conductor, yet it is possible to hold one end of a tolerably long poker in the hand while the other end is red hot. If, however, the poker were of silver no one could touch it. Silver is so good a conductor of heat that it excels all other conductors in this respect. A silver spoon left for a minute in a hot cup of tea burns one's fingers, where a plated spoon, that is, one of commoner metal merely coated with silver, would remain comparatively cool. A silver teapot quickly takes the temperature of the hot tea which it contains, and if the silver were continuous, the handle would be too hot to hold; so that silver teapots are always made with a little strip of ivory on each side of the handle to cut off the conduction of heat.

Radiation.—There is another very important way in which heat passes from one place to another. When the sun shines and makes us feel hot, the heat travels all the long way from the sun to the earth, but it does not come by convection nor yet by conduction; there are no heated currents set up which flow in our direction, nor is it handed on from particle to particle through the air. Indeed, the air may often be very cold even when the sun's rays are passing through it to warm the earth, and anyway, there is no air at all beyond a few miles from the earth. We will only mention here that the process by which it travels is called *radiation*, but the study of radiation and radiant heat will not come within the compass of these lessons. Suffice it to say that radiation is a wave-motion, transmitted through an extremely elastic and subtle medium, that fills all space and pervades all substances, called the *luminiferous ether.** This "light-bearing" ether is, in fact, the same medium that transmits light to our eyes, and radiant heat may be called invisible light; for it travels with the enormous velocity of light, and obeys all the laws of light, some of which we must proceed to study in the next chapter.

* The young reader must not confound this word with the liquid ether sold by chemists. The luminiferous ether cannot be weighed, nor detected in any way by the senses, nor removed from any space; its existence is assumed for good reasons, by scientific men, and its properties are only inferred.

CHAPTER XVII.

LIGHT.

Now, let us learn a little about light.

If we go at night into a dark room we can see nothing at all; but if we light a candle we can instantly see the things in the room. What is it then that happens when the candle is lighted? We will not ask at first what causes the light, nor what the light is; but only *where* it is. It is in the candle flame certainly; but instead of staying only in the flame the light pours out in all directions. Some of it comes to our eyes, and then we see the candle; the light, whatever it is, brings us the knowledge of the candle though it is at a distance from us.

But there is a book lying on the table by the candle. We can see the book too; does light come from that to our eyes to bring us the knowledge of it? Yes—until the candle was lighted nothing came from the book; but some of the light that comes from the candle strikes against the book, and then, bounding back from it, is scattered in all directions; some of this scattered light comes to our eyes, bringing with it the knowledge, or the picture, of the book. We say the light is scattered, or irregularly *reflected*, from the book. Though it came out first from the candle, this light does not bring the picture of the candle, but of the book, which was the last thing it struck against.

Suppose we take the candle out-of-doors on a still dark night to light us in walking. As long as we walk with it between houses and walls, some of the candle-light is thrown back from them, and though it is not very bright, yet we can see pretty well, and there seems to be some light round us; if we walk beyond the houses between trees and hedges they will still throw back a little light; but if these come to an end, and we walk over a bare common or down, it becomes much darker, and we feel as if we and the candle were alone in the dark. Of the light now around us we only know of a little which comes straight to us, making us see the candle, and of a little which is reflected from the ground; all the rest pours away into the darkness. But if some one meets us and stands near us, there will again be something to send back light, and it will be less dark, especially if the person is dressed in grey or white clothes. There was just the same quantity of light coming out of the candle all the time, but we could see very little until the light fell on something which could scatter the rays back to our eyes.

Luminous Bodies.—The candle flame is what we call self-luminous; that is, it gives out the light by which it is seen. The sun is a luminous body; so are most of the stars, and the electric spark, and red hot coals and some other things—we see them by their own light. Everything that can be seen at all must either give out light of its own, or else, like the moon, reflect light received from some luminous body, and if there are any substances that neither give out light nor reflect light, then they can never be seen—they are invisible.

Transparency, Translucency, Opacity.—Take a piece of clear glass and hold it up between your eyes and the candle. You still see the candle and the flame quite distinctly, almost as if there were nothing between it and you. The light, whatever it is, comes right through the glass.

Now, put the flame inside a lamp shade of ground glass. What can we see? The light is still bright; it comes through the globe and falls brightly on all the objects round, only we cannot see the actual flame through the globe.

Take off the globe and put the candle into a dark lantern, closing the shutter. The light disappears. It does not pass through the tin case, and we can only see traces of it at chinks or openings.

Substances like the clear glass, which allow us to see the forms of things through them, are called *transparent;* those like the ground glass, through which we can see light, but cannot distinguish forms, are called *translucent;* while those which, like the tin, shut out light altogether, are said to be *opaque.*

The most transparent substance that we know is clear air. It lets all the light pass right through, and does not reflect any of it ; so that, as it neither gives out light nor reflects light, it is invisible. Clear glass, crystal, and water, are also transparent, but not perfectly so. They do reflect a little of the light from their own surfaces, and as this reflected light brings a picture of them they are not quite invisible, though the greater part of the light passes through them, carrying on the pictures of the objects from which it came before. So, then, by the light which they turn back and reflect, we can see *them*, and by the light which they allow to pass we can see *through* them.

We can find substances of many different degrees of transparency and translucency; different kinds of glass, white or coloured, oils and other liquids, oiled paper, celluloid, etc.; and some we may call half-translucent, like rich-coloured jewels, which seem to let us see a little way into them, though not right through them. Many things also are transparent if we can get them in thin enough slices, while thicker plates of the same are only translucent, and still greater thicknesses are quite opaque.

Reflected Light.—We saw just now that light falling on glass was divided, part of it being turned back or reflected from the surface and part passing through; and light always is divided, at the surface of every visible object. When the candle-light falls on a book on the table, the book does not reflect all the light it receives; it reflects some, or it could not be seen. If a piece of glass lies on the table beside it, this also reflects some of the light so that it can be seen; but a good deal passes through to the table beneath it and lights that up too. In the case of the glass the part of the light that is not reflected is said to be *transmitted*, or passed through; but in the book, and in all opaque objects, that part of the light that is not reflected is said to be *absorbed*—it is lost to sight.

Now, try and observe more closely and carefully still, and we shall find that even the reflected part of the light is divided in two again. Lay a common red pencil on the table where the light can fall brightly on it, and looking at it attentively you will see a narrow line of white light along its whole length. If you do not see this well at first, move the pencil about a little till you find a position where the line is quite bright and sharp. On each side of this white line stretches the red surface of the pencil. Here you see that the light which falls upon the pencil is reflected by it in two different ways— a narrow strip of white light and a wider band of red light.

However solid any object may seem to us to be, its minute particles or molecules are not really quite close together; but there is a little, a very little, distance between them, and each particle has surfaces of its own. When the light first strikes the outer surface of an object part of it glances off immediately, being reflected as white light; and the smoother and more polished the surface, the brighter and more dazzling is this surface reflexion, as we may see if a silver spoon or a looking-glass is set in the sunshine. The white line on the pencil is this first or surface reflexion, which painters call the high light.

But part of the light gets in among the particles, and is reflected backwards and forwards among their innumerable surfaces, and here it goes through a sort of sifting before it comes back to our eyes. Let us try to understand this.

Colour.—What we call white light is really compound, made up of the many brilliant coloured rays which we see in a rainbow, and the particles of objects select and choose among these coloured rays, reflecting some and absorbing others. Our pencil appears red because when the light enters among its particles they absorb all the other colours and reflect only the red. Roses and strawberries look red for the same reason; but in their leaves, on the other hand, the red rays are absorbed and green light is returned to our eyes. And in the same manner everything is at work dividing up the rays of light, and selecting some to reflect and some to absorb. A white object reflects an equal proportion of all the colours in light; so do grey ones, but the whole amount of light reflected by them is smaller, becoming less and less as the grey gets darker, until at last a perfectly black object, in its blackest or shadowed part, reflects none of the colours.

The red rays reach us at the same moment from every part of the space occupied by the pencil, and tell us that there is a long, narrow, straight, red thing on the table. And the rays also give several other particulars about it. The red light being much darker on one side, and getting gradually lighter to the other side, makes us perceive that the object is rounded; and finally the brightness and sharp definition of the white line tell us that it is also polished. The part of the pencil that has been cut is not polished; the white line is not to be seen there, though it may probably be found again on the black lead point.

Now, this is what we mean by the light bringing us a

picture of the object which reflects it. Bundles or sheaves of rays come from every part of the object, so showing its shape; the choice of the particles among the coloured rays shows us its colour; and delicate little variations of light and shade and brightness in different parts give us all details about its surface. The picture of an object, therefore, is only brought by the rays that have entered among its particles, while the mere white surface reflexions pass on, having indeed the directions in which they are travelling altered by the reflecting surface, but still carrying with them the picture of the thing from which they came before.

We can see this if we put beside the pencil a smooth glass ink-bottle with ink in it, taking care that its outside is clean and bright. Its high light, or surface reflexion, is a distinct white spot, which proves, on looking closely, to be a tiny picture of the window through which the light comes. The surface reflexion, you see, brings no picture of the ink-bottle, but only carries on the picture of the window; and it is much more distinct on the ink-bottle than on the pencil, because the glass is more highly polished.

In the same way it is by means of the first or surface reflexion that we see our image in a looking-glass, while the second reflexion, when the light has entered among the particles, enables us to see the mirror itself.

The surface reflexions (which are best seen in mirrors) always follow a certain regular plan as to the directions in which they travel.

Regular Reflexion.—Stand right in front of a looking-glass, and you can see yourself in it; the light carrying the image of your face strikes straight upon the glass, which reflects it straight back to your eyes; but move a few steps to one side, and you can no longer see yourself, but only things on the other side of the room. Look at this figure of the way in which the light is reflected. The line *mn* is meant to represent a section or strip of a mirror. When you stand right in front of the mirror it is as if you stood at *d;* then the rays go straight from your face to the mirror and come straight back; but when you move to one side, as to *b*, then the rays carrying the picture of your face go to the mirror at

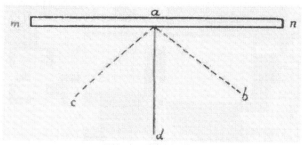

Diagram of Reflexion.

a, and from there bound away to the other side, and are reflected towards *c*, while the light from things at *c* bounds off the mirror towards you, so that what you see in the mirror are all the things in the direction of *c*. Notice that the lines from *a* to *b*, and from *a* to *c*, are just the same inclination to the line *ad*, only on opposite sides of it. The picture would serve as well for an illustration of tennis balls rebounding from a wall as for rays of light, for a ball hit from *b* to *a* would bound off towards *c*, just as the light does.

We will make an experiment to show the path of the

reflected rays. Take a box with straight sides and a glass lid, paint it all black inside, and lay it on its side so that the glass becomes the front wall. Or it will do as well to lay on its side a common box without a lid, and put a pane of

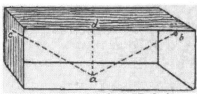

Experiment to show law of Reflexion.

glass to close in the front. Lay a flat piece of looking-glass, or of smooth bright metal, in the bottom of the box as it lies on the table, at *a*, in the picture, and make a small hole near the top of one end, as at *b*, for a sunbeam to shine in through. If you hold the box in the sunshine so that a sunbeam can come through *b* and fall upon the mirror, it will be reflected up again towards the other side of the box, and if the box is arranged so that the sunbeam strikes the mirror just in the middle, as at *a*, then the bright reflected spot will be at *c*, opposite to *b* and at exactly the same height. As long as there is only clear air in the box the path of the rays cannot be seen, but if we fill the box with smoke, by setting fire to a bit of brown paper and holding it for a moment in the box, which must be quickly closed when the paper is taken out, then the path of the rays will be traced out on the smoke, and we shall see them following the dotted lines on the figure. In the picture there is a line drawn upright, *ad*, from the spot where the light falls on the mirror, and the two lines traced by the ray of light are equally inclined to this central line, only on opposite sides of it.* If we now altered the position

* Both this and the next experiment require to be tried in

of the hole, *b*, we should find the same law holding good. At whatever slant the light falls on the mirror it rebounds at precisely a similar slant. We express this by saying that the angle of reflexion, *dac*, is always equal to the angle of incidence, *dab*, and the two are always on opposite sides of the line *ad*, which we can draw for ourselves perpendicular to the mirror.

Refraction.—Now let us look at the directions of rays of light when they are *transmitted*, or passing into and through transparent substances. Take a good-sized medicine-bottle, with flat sides, and fasten over three sides of it a piece of black stuff to darken it, but leaving a smooth flat side uncovered to look in through, as here shown. At *a* cut in the black cover a narrow slit with sharp edges. Fill the bottle about half full of water in which a few drops

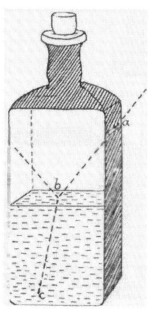

Experiment to show Refraction.

of milk are mixed so as to make it a little cloudy. Light a paper match, and hold it in the upper part of the bottle for a moment so as to fill it with smoke ; then withdraw it quickly and cork down the bottle tightly to keep in the smoke.

If it is now placed in the sunshine so that a sunbeam

sunlight. Lamplight will not serve unless special arrangements are made to correct its divergence, and also to render it brighter.

can enter at *a* and strike on the surface of the water,* we can see that the beam is there divided in two, a small part being faintly reflected up again, but the greater part passing on into the water. Notice that the sunbeam is not straight, but is broken or bent just where it touches the water, so that its direction in the water is different from its direction in the air. This change of direction is called *refraction*, from a Latin word meaning *to break*, and it takes place when rays of light pass in a slanting direction from one transparent substance into another of different density. Water is denser than air, so that the beam is bent or refracted in passing from air to water.

There is only one position in which light would not be refracted in going into the water, and that is if it came down the neck of the bottle and struck straight on the surface of the water; the beam would then be *perpendicular*, or *at right angles*, to the surface of the water, and in this position it would pass straight into the water without changing its direction.

It is so important to understand clearly the meaning of *perpendicular to the surface* that we will illustrate it a little further. Set the flat end of a pencil against a mirror, and look at the pencil and its reflexion. If the pencil leans a little to the right the reflexion leans a little to the right too, and the line that they make together is inclined in the middle, as at *b*, on the next page. So it is if the pencil leans in any other direction, as at *c*; but when the pencil and its reflexion, *from whatever point they are looked at*, make an unbroken straight line together, as at *a*, then the

* By means of a little mirror the beam can be made to enter at *a* without tilting the bottle.

pencil is perpendicular to the surface of the mirror. When the mirror is lying quite flat the pencil will be standing quite upright; but whatever position the mirror may be in, the perpendicular to its surface can always be found by the same method. You see there is only one direction in which the pencil can point when it is perpendicular to the surface; if (with one end still touching the mirror) it points in any other direction, it is said to be *oblique* to the surface—less oblique when it is not far from the perpendicular position, more oblique if it leans

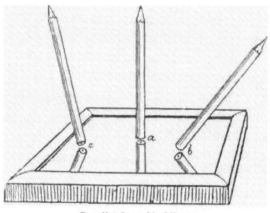

Pencil reflected in Mirror.

more to one side, as at *b*, and very oblique when it is nearly lying down on the mirror. Look back at p. 267, and you will see that the sunbeam that entered the bottle obliquely to the surface of the water, became less oblique or more perpendicular to the surface after it had passed into the water. Light is always bent more towards the perpendicular direction in the denser substance.

Apparent Place of Objects.—We come now to a very important point which must be carefully remembered.

Light, we know, carries with it pictures or images of objects; but if the rays of light bringing these images to our eyes have rebounded, or been bent during their journey (either by reflexion or by refraction), where shall we see the image? *The image always appears to be in the direction from which the rays came last.*

Stand again in front of the looking-glass, *mn*, not right in the middle, but as at *c*, in the picture, and put a candle

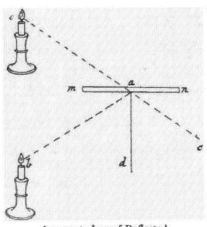

Apparent place of Reflected Object.

at *b*, and an image of the candle will be seen at *e*, apparently behind the glass, or, as we say, *in* the glass. The rays which produce this image have gone first from *b* to *a*, and then from *a* to *c*, so that to any one standing at *c* the image is seen along the line *ca*, that is, in the direction from which the rays came last. In fact, the image appears to be as far behind the mirror as the object really is in front of it.

Just in the same way when the rays are bent by refraction, we see the image in the direction from which the *refracted* rays reach us. Place a shilling in the bottom of a glass basin and pour in water gently so as not to move the shilling. When looked at obliquely from above, the apparent position of the shilling will rise higher and higher as the water rises in the basin. It will still appear to rest on the bottom, because the apparent place of the bottom of the basin is equally raised by the refraction; but if the eye is so placed that the edge of the basin, when empty, just hides the shilling, the appearance of the shilling will rise into sight on pouring in the water. We have seen that a ray falling obliquely on water, as *a b* (see picture), will be refracted towards *c*, but it is equally true that if there is an object at *c*, some of the rays bringing its picture, will travel up through the water to *b*, and will there be refracted in the direction *b a*. If these rays are received by an eye at *a*, where will the object at *c* appear

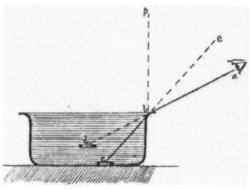

Apparent place of Coin altered by Refraction.

to be? It will appear to be *in the direction along which the rays came last*, and consequently will be seen along the line *ab*, at about *d*. As long as there was no water in the basin, the shilling at *c* could not be seen because the edge of the basin cuts off the direct rays from *c* to *a*, but the addition of the water has enabled us to see over the edge, without raising the eye to *c*.

For the same reason a stick plunged obliquely into water appears bent. The rays bringing the image of the end of the stick that is under water (that is, at *c* in the picture below) travel in the direction *c b;* at *b* they are refracted into the direction *b a,* so that the eye at *a* sees the end of the stick along the line *a b,* and the point of it appears to be at *d,* and so with every other portion of the stick that is under water. Similar reasoning explains how every portion of the stick that is under water (from *c* to *c*) appears displaced into the position

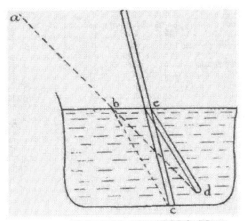

Refracted Image of Stick in Water.

c d, when viewed by the eye at *a.* In like manner the apparent place of the bottom of the vessel is also raised.

Proportion of Light reflected.—In our waterbottle (p. 247), we saw that very little of the light was reflected from the surface of the water, most of it being refracted; but the quantity of light reflected or refracted depends on the direction in which the light strikes the water. Place a sheet of smooth, white writing-paper on the table between you and a candle, and gradually raising it to the height of your eye, notice how much brighter the reflected light becomes. When looked at very obliquely, an image of the flame may be seen reflected in the paper. So when light falls perpendicularly upon water only one-fiftieth part of it is reflected, but as the beam falls more obliquely more and more of it is reflected, till the reflected part is equal to three-quarters of the whole beam. Just the same kind of variation takes place with other reflecting surfaces.

Effects of Refraction.—It would be very interesting to show the effects produced by causing beams

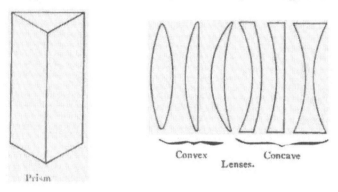

Convex   Concave

Lenses.

Prism

of light to pass through transparent substances of certain definite shapes. We should see, in one case, that ordinary white light passing through a three-sided piece of glass, called a *prism*, may emerge divided into all its beautiful colours, the colours of the rainbow; while, by the passage of light through differently shaped and arranged *lenses*, objects can be made to look larger or smaller, farther off or nearer than they really are. All these effects are produced by refraction.

The next diagram shows how rays of light, starting from a point, *f*, say a candle flame, diverge in all directions, and then by the "double convex" lens are made to converge at a point, *f'*. The converse would take place if the candle were placed at *f'*. The two points *f* and *f'* are called *conjugate foci* of the lens. If the rays fall on the lens from a very distant source of light, such as the sun, the focus to which the rays would be bent

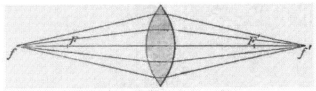

Refraction of light by a convex lens.

is then called the *principal focus*, as the rays falling on the lens from a very distant light, are practically parallel. If now the source of light is put at the place where the principal focus, F, would be, then the rays passing through the lens emerge parallel to one another. If the candle be placed between *f* and F, the rays, after passing through the lens, would converge to some point more distant from the lens than *f'*.

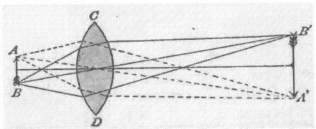

Formation of an enlarged inverted image by a double convex lens.

In the case of a magic lantern the slide or object is placed in this last position, as at A B in this picture; there is thus formed, after the rays have traversed the lens C D, a distant, enlarged and inverted picture, A′ B′. The rays from the point B are, you see, so refracted as to form an image at B′, the rays from A likewise form an image at A′, and so on with all the intervening points of the object A B. To get the picture the right way up, the lantern slide, corresponding to A B, must therefore be upside down. If the object were at A′ B′ the picture, or *image*, would be at A B; this is similar to what occurs in a *camera obscura*, or a photographic camera, or the eye, a smaller inverted image of outside objects is seen at A B, where the photographic plate, or the retina in the case of the eye, is placed.

Velocity of Light.—The only other point that can be mentioned here is the enormous speed at which Light travels—a rate of not less than 186,000 miles a second. Here on the surface of the earth it is practically instantaneous. We can see the light of a match the instant it is struck; not an instant is wasted while the light is passing from the spark to our eyes.

But though light travels at this extraordinary speed, yet the distances of the heavenly bodies from us are so great that light actually takes eight and a quarter minutes in reaching us from the sun; about five hours in coming from the planet Neptune; not less than 3½ years from the nearest fixed star; and probably centuries in coming from the nearest nebulæ.

Light is the only messenger conveying to our bodily senses communications from these vast distances, so that our knowledge of the stars depends entirely on the information which it brings.

CHAPTER XVIII.

SOUND.

WHILE Light in its swift journeys brings us intelligence alike from far and near, we have, with our nearer surroundings only, another means of communication in Sound, the sound of voices, of music, and of all kinds of noise; and a thunderstorm gives us a good opportunity of comparing the relative speed of Light and Sound.

Velocity of Sound.—When there is a storm close to us we see the lightning flash and hear the thunder clap at the same moment; but when the storm is further off we see the lightning first, and then have to wait for the sound of the thunder. What is the reason for this difference? The lightning and thunder really take place at the same time; but while the Light reaches us instantly, travelling at the rate of 186,000 miles in a second, the Sound lags behind, and we have to wait while it is coming. Through air at an ordinary temperature it only comes at the rate of 1120 feet a second, and by counting the number of seconds which elapse before it arrives, we can tell how many times 1120 feet the sound has travelled, and so ascertain the distance of the storm.

In the same way, if we watch any one firing a gun some way off, we can see the flash and the smoke before the report reaches us; and a quaint effect is produced by regular hammering or wood-chopping at a little distance, the sight and the sound of the blows coming not together but by turns.

But the velocity of sound varies according to the substance through which it is passing. Through water it comes four times as fast as through air, and through most solids faster still, the speed in iron being about fifteen times as great as in air. Put your ear against a

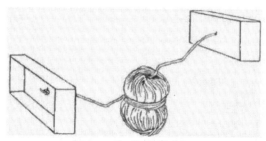

A toy mechanical telephone.

telegraph-post while some one hits the next post with a stick, and you will hear the sound of the blow twice, the first report running very quickly through the telegraph wire and the wood, while the second is travelling more slowly straight through the air.

It is easy to make a toy telephone which will carry sound a long way. Take two empty boxes of thin wood without covers (cigar boxes do well), bore a hole in the bottom of each, and pass through each of them one end of a long string, or still better, of a wire, say a long bell-wire; now knot the ends, and take the boxes as far

282

apart as the wire will allow.* The wire should be kept tightly stretched, but it need not be kept straight all the way; you can bend it round a corner without any harm. Turn the bottoms of the boxes towards each other; and then if you place your ear against one of them you can hear any tap that is made upon the other, by the sound travelling through the wire. If a tuning-fork is struck and made to sound near one box it will be distinctly heard through to the other, and if you whisper into one box your whisper will be heard at the other. Neither tuning-fork nor whisper can be heard through the air, if your distance apart is sufficient, because the sound waves spread in all directions through the air (unless confined by a speaking tube), and so their loudness rapidly fades away. The wire, or stretched twine, conveys or *conducts* the sound along itself, and so prevents its decay at moderate distances. This is the principle of the *mechanical telephone* which the writer has used for conversation between two houses a mile apart, even the faintest whisper at one end being heard at the other.

Connexion of Sound and Motion.—Whatever is producing sound is in motion. Strike a tuning-fork sharply and hold it in your hand, and you can feel its motion all the time the sound continues. Sometimes we can actually see the ends moving, and by the following simple plan we can certainly show the motion.

With a little morsel of wax fasten on to each prong of a tuning-fork a small bright silvered bead. (One would

---

* The picture shows a ball of twine that is intended to be stretched tightly between each box; but you will find it easier to take a shorter length than this. The *electric* telephone is of course a different thing.

do, but something to balance it must be put on the other that the weight may be the same.) Rest the end of the fork on the table where the sun can shine on it, and notice where the bright reflexion falls from one of the beads; if it falls in a shadowed place it will be more visible. Then strike the fork and rest it as before, and you will see the spot of reflected light lengthened out into a bright line by the rapid shaking or vibrating of the bead backwards and forwards.

The following experiment is perhaps easier for you to manage. Instead of fastening the bead to the fork, hang up the bead by a piece of thread; or a little ball of sealing-wax will do as well as the bead; now strike the tuning-fork and bring it carefully to the bead; directly the tip of the fork touches the bead, away the latter is dashed, and as it falls back gets another blow from the vibrating fork, and so on till the tuning-fork ceases to sound, and then the bead hangs motionless.

Here is another pretty experiment. Fill a round glass finger-bowl about half full of water, and set it on a tablecloth or something soft. Wet your finger and draw it with a firm but light pressure round and round the rim, until the glass gives out a clear musical note. Continue the movement, and presently you may see a corresponding movement in the water, which rises in a tiny wave of exquisite ripplings, and follows the course of the finger round the glass. Here the glass is thrown into rapid vibration (or movement of its substance backwards and forwards), and the motion being communicated to the water shows itself in visible waves.

These have been examples of small, rapid, delicate

movements with rather soft sounds; but very loud sounds imply much more violent movements. Any one who has heard the deep pedal notes of a large and powerful church organ knows how they will sometimes shake the whole building, and the hearer feels them as a great throbbing all through him; while the violent shaking of the air that is produced by cannon firing or by blasting is sometimes sufficient to break windows in the neighbourhood.

Vibratory Motion.—We saw that the ends of the tuning-fork were moving rapidly backwards and forwards all the time it sounded, that the glass of the finger-bowl and the rippling water were vibrating to and fro, that the organ notes made a continuous throbbing. Even the cannon firing and the blasting do not end in single motions, but set the air rocking violently. And so it always is. The motion which produces sound is never a single motion, but always consists of vibration or swinging backwards and forwards of the sounding body; not an imperceptible motion of its invisible particles or molecules, as is the case with a luminous or a hot body, but a vibration you can feel, or that can be made perceptible by simple means. Moreover, the sounding body vibrates as a whole, or in a few aliquot parts, comparatively slowly (see p. 282), whereas the molecules of a luminous or a hot body vibrate with inconceivable rapidity, millions of millions of times in every second, and set going waves *in the ether* (p. 257), which come to us as light or radiant heat. The sounding body produces waves *in the air*, which travel vastly slower (p. 276), and on reaching our ears give rise to the sensation of sound.

But when there is nothing to carry on the waves sound does not travel at all. It can pass along solids, as we found with the telegraph-post and the toy telephone ; it is transmitted by liquids—a man floating with his ears under water can hear sounds—and we know that it comes through the gases of the air ; but in a vacuum, where there is neither solid, liquid, nor gaseous substance to carry it on, all sound dies out.

Here is a diagram of an air-pump. You see that the piston (*a*), working up and down in a cylinder, is arranged exactly like that of the water-pump on p. 227 ; but the tube below (*b*), instead of dipping down into water, passes into a bell glass (*c*), closely fitted upon a plate ; so that by working the piston up and down we can pump all the air out of the glass. If a watch, or, still better, a small, loud-ticking American clock, is placed under

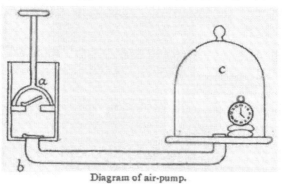

Diagram of air-pump.

the bell glass (resting on a thick pad of cotton wool to prevent the plate carrying the vibrations), we shall hear it through the glass ticking away merrily. But on working the piston so as to pump away the air, the ticking can no longer be heard ; sound cannot cross a vacuum.

286

You will have noticed that when a tuning-fork is held in the air very little sound is heard, but when set on a table the whole table-top vibrates with it and the sound is much louder, because a much larger surface is set in vibration. So a violin string stretched in the air and set vibrating gives but a very weak sound, but when stretched upon the thin hollow violin, both the front and back of the case and all the air inside it vibrate together and give out a powerful sound.

When vibrations are slow, they are merely separate throbs; but when they follow each other with not less than a certain rate of speed, they become audible as a continuous sound.

Musical Sounds.—If the vibrations are irregular, we hear merely a noise ; but if they are regular—that is, follow each other at regular intervals—and at the same time are sufficiently rapid, we have a musical note. The pitch or height of the note depends on the number of vibrations in a second. Slower vibrations give a deep, low note : when they come faster the note rises in pitch, and very rapid vibrations cause a high shrill note. Faster still they cannot be heard at all !

Sixteen vibrations in a second make a note about as low as can be heard at all, and by the word vibration, we mean a complete swing to and fro ; double that number, 32, give a note an octave higher; 64 is an octave higher again, and so the notes rise, doubling the number of vibrations with every octave, until we reach the acutest, shrillest squeak that can be heard at somewhere about 30,000 vibrations in a second, making in all some eleven octaves. There is, however, a good deal of question as to the extreme limits at the top and bottom of the range of hearing, and there can be no doubt that people differ a good deal in their actual power of hearing very deep or very acute sounds.

When old age is coming on, and the hearing becomes a little dull, the first failure is in the power of hearing very shrill sounds. A short time since, a party of travellers were riding on mules over the pass of the Grimsel, in Switzerland, and an elderly gentleman remarked how weary he was of the dull thud, thud, of the mules' feet upon the turf; but his younger companions could not hear it on account of the shrill noisy chirruping kept up by myriads of large grasshoppers beside the way. This noise, though so loud to them as entirely to conceal the sound of the mules' tread, was absolutely unheard by the old gentleman, who, with advancing age, had lost his former power of hearing sounds so high in the scale.

But whilst a good ear can distinguish sounds through a range of some 11 octaves, no one would call the lowest and highest sounds *musical;* in fact, the sounds which are musically available only range from about 40 to 4000 vibrations a second, that is, about 7 octaves.

The middle C.

The note called the middle C on the pianoforte here shown has about 260 vibrations in a second, though it is sometimes placed a little higher or lower according to different standards of pitch. This is a useful central note to remember. To compare the notes of a scale any tone may be chosen as a keynote. If we take G, then the keynote has 384 vibrations in a second, the third note above it has 480, the third above that again, or the fifth of the scale, has 576, and the octave, as we know, double the keynote, or 768. These four notes, keynote, third, and fifth, completed with the octave above, make what is called the common chord of the key here shown, and whatever keynote may be taken, the number of vibrations in the notes of its common chord always have this same fixed proportion to one another; the third a

The common chord of G.

quarter more than the keynote, the fifth a half more, and the octave double the keynote; this proportion is as the numbers 4, 5, 6, 8.*

There is no prettier or more interesting subject of study than the close relation between music and mathematics.

* Taking the numbers given for the keynote of G, namely, 384, 480, 576, and 768, and dividing each number by 96, you will find the proportion of 4, 5, 6, 8, as stated.

CHAPTER XIX.

MAGNETISM.

IF you go to a toy-shop and ask for a magnet (which may be bought for sixpence, or one shilling) you will receive a piece of steel bent into the shape of a long horse-shoe. There will be a second small piece of iron with it, bridging across the opening between the two ends. Pull off this piece (which is called the *keeper*), place it on the table, and bring the ends of the horseshoe piece near it, when it will be attracted, and will stick quite firmly to it, requiring some considerable force to pull it away. Take it off again, and put on the table other small pieces of iron or steel—nails, needles, or anything you like—and you will find them all attracted in the same way. The horse-shoe, dipped into a box of small nails, will come out bristling, like a hedgehog, with nails sticking all over its ends. Any piece of steel like this, which will attract other pieces of iron, is called a magnet, and the property by means of which it acts is called *magnetism*.

Notice that the attraction of a magnet is not of the same kind as that exercised by an excited glass or wax rod, which we shall study presently under the head of electricity, for, while those will attract any light insulated object for a moment, repelling it again as soon as it is touched, the magnet attracts only pieces of iron, and they remain sticking firmly to it. Try to attract bits of paper or bran by your magnet and you will not succeed. Besides, the attractive power of a rubbed glass rod will disappear at once if it is not insulated, but we can handle a magnet freely without affecting its power of attraction.

Take a short steel knitting-needle, and, laying it on the table, stroke it several times from end to end with one end of the horse-shoe magnet. It does not matter which end you use, but you must keep to the same throughout, and must always stroke the needle in the same direction, never rubbing backwards and forwards. After this treatment you will find that the knitting-needle has also become a magnet, and will attract small pieces of iron. A second, and, in some ways, more convenient method of making the needle magnetic, is this: hold it in your hand, and stroke several times from the middle to one end with *one* end of the magnet, and then, turning it round, stroke from the middle to the other end with the *other* end of the magnet.

Now, lay your magnetic needle in a loop of paper, pass a thread through the paper, and knotting its ends together, hang it up to a support, so that the paper and needle can swing freely. Notice that the needle takes up a certain position ; we may push the end round to one side, but it will return to the same position : we may move the support round, but the needle, after swinging a little, will always settle down in the same direction as before, and this direction is very nearly due north and south. So, then, we have found out that one end of a magnetic needle, if free to move, will turn and point to the north and the other to the south. One end is called the North-seeking pole, or, for short, the North Pole, while the other is the South-seeking or South Pole. Mark your magnet in some way, so that you may know

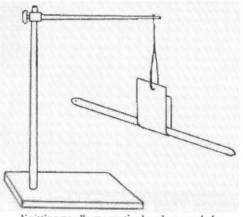

Knitting-needle magnetized and suspended.

its north pole from its south; a dab of paint will do, or a tiny morsel of paper gummed on.

Now, let us magnetize a second knitting-needle, and after testing and marking it in the same way, we will balance it in the paper loop, and then bring the north pole of the first needle near to the north pole of the second. Notice that the end of the needle moves away from it. Bring the south poles near each other, and we find again that the south pole of one magnet moves away from the south pole of the other. But if we present the north pole of one to the south pole of the other, we find that they attract each other. From this we see that *like poles repel* one another, and *unlike poles attract;* a result which, as we shall see presently, reminds us of the repulsion and attraction between like and unlike charges of electricity.

Here is a pretty experiment. Lay the magnets parallel to each other a little way apart on the table, place a sheet of stiff white paper over them, and then sift down upon it some fine iron-filings through a small fine muslin bag. See how curiously the filings arrange themselves in curved lines; they feel and obey the attraction of the

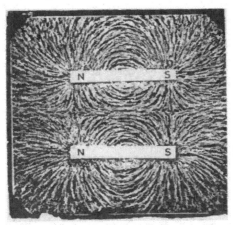

Magnetic curves, similar poles juxtaposed.

magnets through the paper. If we put the south pole of one magnet opposite the south pole of the other, at a distance apart of about an inch, two sets of curves are formed which seem to be independent of one another; but if a north pole is placed opposite a south pole, the curves pass from one magnet to the other. These curved lines are called *lines of magnetic force,* and the space through which they are spread is called *the magnetic field.* If you like to make these curves permanent, instead of the plain white paper lay down a sheet that has been gummed all over and allowed to dry; then when the filings have arranged themselves, without moving the paper, let some steam pass *very* gently over the surface (if the gum is fairly thick even breathing over it may be enough), and the gum will soften for a moment and dry again, holding the filings in their places.

We know that a magnet has a north and a south pole.

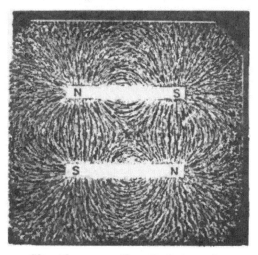

Magnetic curves, unlike poles juxtaposed.

Now, what would happen if we broke it in two through the middle? Should we have one piece all north pole, and the other all south? No; we find that each piece is a complete magnet, the broken ends becoming a south pole to the north pole half, and a north pole to the other half. No matter into how small fragments a magnet may be broken, each will still have north and south poles of its own and be a complete magnet. It is impossible to separate the poles of a magnet.

We have converted small steel bars into magnets by merely stroking them with a magnet, and now that we know the difference between north and south poles we may notice that if we stroke with the north pole of the magnet, the end *last* touched of the new magnet will be a south pole; or if we stroke with a south pole, the end last touched will be a north pole.

But try the same experiment with a key or an iron nail (made of so-called "soft" iron). Stroke carefully and then dip either point into the iron filings; they do not stick to it. Touch the head of the nail or key with a strong magnet, and they will stick to it instantly; whilst thus sticking on, dip the lower end of the nail into iron filings; now they fly to the nail, which has become magnetized by being near to the magnet, but as soon as the magnet is removed from the nail the filings will drop off again. So it appears that *hard steel* becomes a permanent magnet when properly stroked with a magnet, but that *soft iron* cannot be permanently magnetized in this way; it is a magnet while the steel magnet is touching it, but when that is taken away all the magnetic power disappears.

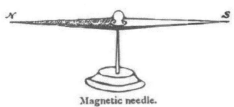
Magnetic needle.

There is another way in which a piece of soft iron may be converted into a temporary magnet, which is by the use of electricity; but we shall come to this presently.

The most important use of the small magnet, or magnetic needle, is in the mariner's compass. In this a magnetic needle is balanced on an upright point, so that it can swing round freely; and as it places itself pointing nearly north and south, the sailor can tell from it in what direction he is steering his ship.* Before the discovery of the compass, ships could rarely venture to sail out of sight of land, for though they might be steered by the sun and stars, yet they were in great danger of going astray when these were hidden by thick clouds.

Magnetism of the earth.—We said just now the compass needle points *nearly* north and south. Why is this? The earth behaves as if it were a huge magnet, with its poles near to, but not coinciding with, the geographical poles. As it is to the magnetic pole the needle points, this want of coincidence gives rise to what is called the *variation* of the compass. The north magnetic pole is in the Arctic region of Boothia, a place north of Hudson's Bay; the south magnetic pole has not yet been reached. Now look at a map, and you will see that in Europe and the Atlantic the needle must point west of the true geographical north, whilst in the Pacific, etc., it will point east of the true north, and on some meridian between, the compass will have no variation or point true north. Thus the variation of the compass is different, both in amount and direction, in different longitudes on the earth's surface.

---

* In the mariner's compass a card, called the compass card, is fixed to the needle and moves with it. This card has its rim divided into 32 equal parts, called the *points* of the compass. The vessel is steered by keeping a certain line, corresponding to the head of the vessel, opposite the right point on the compass.

CHAPTER XX.

ELECTRICITY.

EVERY one has heard of the wonders of Electricity—of the Telegraph, by which messages can be sent hundreds of miles in shorter time than it takes to say the words; of the Telephone, which enables a man to speak to another hundreds of miles away, and yet be heard as distinctly as if he were talking only through a short length of pipe; of the bright Electric Light; and of trains, tramcars, and other vehicles, worked by Electricity.

What this powerful agency *is* we know not. We are as ignorant of its actual nature as we are of the nature of Gravitation, the falling together tendency; and even to explain fully the way it works would require volumes. All we can do here is to learn by simple experiments some of the first things which *must* be known before we can possibly understand more.

FRICTIONAL ELECTRICITY.

Let us recur to the little experiment mentioned on p. 205. Tear up some thin light paper into pieces about half the size of a threepenny piece; then take a stick of sealing-wax, or some article made of black ebonite, or, if you like, the amber mouthpiece of a tobacco-pipe, rub it well on your sleeve or some other woollen substance, bring the end of it near the pieces of paper, and you will see them jump up, touch the rubbed body, fall off, jump up again, and so on for a considerable time. Now, instead of the sealing-wax try a glass rod rubbed with a *silk* handkerchief, first making both very dry by holding them in front of the fire until they are quite hot, and we shall find that the glass rod has the same power of attracting the bits of paper.

Sealing-wax and glass rods in their usual state have no effect of this kind, so that it is plain that some difference has been made in them by the rubbing. We say that they are *electrified*, or electrically excited, or charged with Electricity. This name is derived from *electron*, the Greek word for amber, for the ancient Greeks knew that a piece of amber when rubbed would attract light bodies, and they thought the rubbing put some sort of life into the amber.

Conduction.—Notice that the rubber and the rubbed body are different substances. So far as we know any two unlike bodies rubbing together become electrified; even the liquid metal mercury shaken in a dry glass gets strongly electrified by its friction with the glass.

At first sight this statement does not appear to be true, for when we take a rod of brass or any metal, rub it, and bring it near the pieces of paper, they are not attracted. If, however, we attach the brass rod to a glass or ebonite handle and rub the brass, holding it by the handle so as not to touch it with the hand, we

298

find that it does now attract the light particles. The reason of this is, that the electricity which is produced on rubbing the metal was able in the first case to pass along the metal to the hand, from the hand into the body and into the ground, and so to escape; while, in the second case, the glass handle prevented it from passing, so that it remained on the metal where it was produced. We see from this that some substances will allow electricity to pass through them while others will not; that there are good and bad conductors of electricity as there are of heat, and that metals are good conductors, while glass, amber, ebonite, etc., are bad or non-conductors. Fastening the brass rod to the non-conducting handle is called *insulating* it, that is, keeping it separate from anything to which it might transmit its electric charge.

Doubleness in Electricity.—We will study a little more closely the electricity produced on different bodies.

Take two very light, small objects, such as tiny balls of pith, thread each on a piece of fine silk thread, and

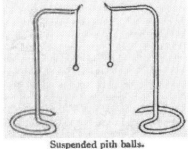

hang them from two pieces of stout wire bent at right angles, bending one end of each wire round into a ring so as to enable the wires to stand upright and the silk threads to hang straight down. The silk threads

Suspended pith balls.

are non-conductors, and their use is to *insulate* the pith balls. Rub the glass rod briskly with a dry silk handkerchief and bring its end near one of the pith balls;

the ball will be attracted and fly towards it. If, how-ever, it touches the rod, it will immediately be repelled and fly away again, and cannot be induced to come near a second time. Rub the sealing-wax sharply with flannel so as to electrify it, and try its effect. The ball which was flying from the glass is strongly attracted to the wax. If we bring the sealing-wax near the other ball, we shall find that it will behave as the first ball did with the glass—first attracted by the rod until it touches it, and then instantly flying away repelled. If we follow it with the wax it will keep edging away so as to avoid touching it again, but if we offer it the rubbed glass rod it will fly to that. This experiment may be varied by touching both the pith balls with the same rod ; then, on bringing the balls near together, they repel each other ; but if one be touched with the glass and the other with the wax, they will attract each other.

Mode of suspending rod.

Taking off the pith balls, attach a wire loop to the end of the silk thread and lay the excited wax in it. Then rub a second stick of sealing-wax and present its end to the end of the one suspended in the wire, and they will repel each other vigorously ; but if the excited *glass* rod is held to the suspended wax it will be attracted. In the same way, if we hung up and insulated an electrified glass rod, it would be repelled by another excited glass, but attracted by excited wax.

It is plain from all this that there is something double about the electricity. It cannot be all just the same thing, or why should electrified g'ass attract when electrified wax repels? People talk about two electricities, or two kinds of electricity. It is difficult to know what these words mean when we do not know the nature of electric action; but a doubleness of some kind there certainly is, and it is necessary to distinguish the two things by names. The electricity produced by rubbing glass with silk is therefore called *positive electricity;* and the other sort produced by rubbing sealing-wax with flannel is called *negative electricity.* These names do not mean that one has more electricity and the other less, but that there is a difference between them of such a nature that equal quantities of the two opposite electricities when brought together neutralize each other, and no electrification remains; just as $+4$ and -4, when added together, make o.

We have also learned from our experiments the very important fact that two bodies charged with the same kind of electricity repel one another, while two charged with different kinds attract one another.

By rubbing glass or sealing-wax in the hand, as in the above experiments, only very small quantities of electricity can be produced, but machines have been invented for rapidly producing large quantities. Many of them work on the principle of rubbing glass or ebonite with silk, but the glass is in the form of a cylinder, or of a large circular plate, which is pressed hard against a pad covered with silk, and then turned quickly by a handle.

Electroscopes.—Various instruments have been made by means of which the presence of electricity, even when in very small quantities, may be detected : such instruments are called *Electroscopes*. The pith balls, or the little bits of paper in our first experiment, might, in this sense, fairly be called electroscopes, and we may easily find other very simple ones. If we put an egg in an egg-cup, and balance on it a wooden lath about three feet long and as thin as possible, then, on bringing near it a rubbed glass rod, it will turn round and try to touch the rod. Or we may try the same experiment by balancing a straight piece of straw on the point of a needle. So the lath and the straw become electroscopes.

The best form for simple experimental purposes is that known as the gold-leaf electroscope. It consists of a wide-mouthed bottle, through the cork of which is passed a brass rod ; round the brass rod a little shellac has been melted, which prevents it touching the cork, and so insulates the rod. The top of the rod ends in a plate or knob, while to the lower end, that is, the end inside the bottle, are attached two pieces of thin gold-leaf. When the knob at the end of the rod is touched with rubbed glass, electricity passes into the rod, and also, of course, into the gold-leaves, and these being charged with the same kind of electricity repel one another, and stand apart—slightly, if the charge is very small, and more widely when more electricity is present.

Gold-leaf electroscope.

Let us see what more we can learn by means of this instrument. Connect to the rod of the electroscope a long metallic wire; support and insulate the other end of the wire by twisting it round a stick of sealing-wax, which may for convenience be stuck upright on a bit of wood. On touching the further end of the wire with a rubbed glass rod, the gold-leaves will diverge as if the knob itself had been touched. Take off the wire, and replace it with a piece of silk thread, and we now find that an excited glass rod touching the thread has no effect upon the gold-leaves; they do not move or take any notice. So we discover that the silk is a non-conductor of electricity, while the wire is a conductor. But we will dip the silk in water, and try the experiment again. This time the gold-leaves diverge as they did when the wire was used; we have turned the non-conducting silk into a conductor by wetting it, showing that water is a conductor. Now you see why we had to be so careful in our earlier experiments to make everything very dry, for where there is any moisture it will carry away the electricity as fast as it is produced. Try now similar lengths of cotton and linen threads; you will find the linen a better conductor than the cotton, but not nearly so good as the metal wire. Linen, cotton, and string are *imperfect conductors ;* in fact, they do not conduct the ordinary voltaic or current electricity which we shall describe presently.

Electric Distribution.—Warm the end of a stick of sealing-wax in a candle, and stick a penny on to it. If now, holding it by the wax-handle, we touch with the penny any electrified body, part of the electricity will pass to the coin, and, being unable to escape through the non-conducting wax, will remain upon it so that we can carry it away to examine. In fact, the little instrument acts like an electric carrier, with which to carry away a small quantity of any charge ; its proper name is a *carrier* or *proof plane.* We can, of course, *discharge* it, that is, cause the electricity to vanish, by letting it touch a conductor joined to the earth ; the touch of a finger will do it.

Now, suppose we have a metallic ball, and wish to test whether it is charged with electricity. We lay the carrier on it for a moment, and then apply it to the knob of the electroscope, when the presence of electricity will be shown by the diverging of the gold-leaves. But supposing that instead of a ball we have a can, or hollow metal globe, charged with electricity, then the carrier will show us a new fact, *i.e.* that while we can take electricity readily from the *outside* of the can, yet, if the carrier touches the *inside*, and is then applied to the electroscope, the gold-leaves will not diverge, there is no sign of electricity there. Whatever charge of electricity may be given to a vessel, the whole of it is found on the outside, and none whatever inside. This is an important fact to remember.

But though the electricity always shows itself on the outside, it is not always distributed equally over the surface of the electrified body. If, instead of a round ball of metal, we have one in the shape of an egg, and electrify that, then by putting the proof plane on different parts of it and touching the electroscope, we find that the gold-leaves move but little with the electricity from the round end, but fly much further apart when the charge is taken from the pointed end ; showing that most of the electricity resides at the point. In fact, if an electrified body actually ends in a sharp point, so much of the electricity will accumulate on the point that it will escape into the air, and the body will be *discharged*, or emptied of electricity.

Electric Induction.—You will remember that when we brought a key near one pole of a magnet (p. 290), the key became magnetic, and remained so as long as the magnet was near, but its magnetism disappeared directly the permanent magnet was taken away. A similar thing occurs when we bring a conductor near an electrified body; the conductor becomes electrified by the presence

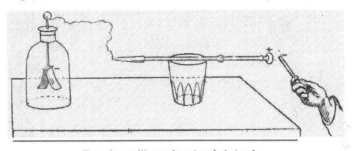

Experiment illustrating electric induction.

of the electric charge near it, but loses its electrification directly we remove the rubbed glass or other electrified body. This influence exerted by an electric charge on bodies in its neighbourhood is called *electric induction*, just as the influence exerted by a magnet on iron is called *magnetic induction*. And the two are very similar; for the north pole of a magnet by its influence produces an opposite or south pole in the end of the iron nearest to it, and repels to the further end of the iron the similar or north magnetism; so, also, a positively electrified body, like rubbed glass, attracts the opposite or negative

305

electricity to the end of any conductor near it, and repels the similar or positive electricity to the far end. Had we used a negatively electrified body, like rubbed sealing-wax, positive electrification would be found on the near side of any adjacent conductor, and negative on the distant side. In every case this induced electrification comes and goes with the approach and removal of the charged body.

Many interesting experiments can be made to illustrate electric induction. Here is one. Support a short metal poker on a dry and warm glass tumbler or other insulator, and connect one end of it to your electroscope by a wire or damp thread (see p. 300). Now bring rubbed sealing-wax, or any electrified body, near to the other end of the poker; notice how the gold-leaves open as the electrified body comes near, and close the moment it is removed. The electric charge induced on the poker is shared by the electroscope which reveals its presence. Next remove the connection with your electroscope, and let your electrified body remain near one end of the poker; now test, by your little carrier and electroscope, the nature of the electrification at each end of the poker. Thus, if you are skilful, you can easily verify the statements in the preceding paragraph.

There is a remarkable difference between the condition of the attracted and the repelled induced electric charge, which can be proved by the same simple apparatus, but these and many other interesting experiments on electric induction you will learn if you pursue the study of *Physics*, of which electricity is one branch. The well-known *Leyden jar*, which enables us to accumulate an electric charge, and the *Electrophorus*, as well as other more modern and powerful "influence machines," all depend upon the principle of electric induction.

Current or Voltaic Electricity.

As yet, in our electric experiments we have been chiefly concerned with electricity remaining quietly on the surface of a body which has been charged with it; now we must find out something about electricity in motion, and for this we shall need a little more apparatus.

Go to a tinsmith's and get a piece of sheet copper and a piece of zinc—a flat plate of each about two inches square, and have a piece of copper wire soldered to each. These must be placed in a jar or glass vessel, taking care that the two plates do not touch each other, and the jar must be filled with dilute sulphuric acid, one part of the acid to about twelve of water. If you now put the ends of the two wires for an instant on to your tongue you will find that they produce a very peculiar taste or feeling.

Now take the compass-needle (p. 290), and wind round it a coil of several turns of "covered" copper wire. Place the compass on the table, and turn it so that the needle lies within and parallel to the coil of wire. If you now touch the ends of the wires from your zinc and copper with the two bared ends of the wire of your coil, you will find the compass-needle instantly driven on one side. If you change over the wires, that is, put the one from the zinc where the copper was, and *vice versâ*, the magnetic-needle will now be driven to the opposite side. Notice that you can keep one wire, say, from the zinc, twisted to the wire of the coil, and no motion of the magnetic-needle is produced until the other wire (that from the copper) touches the other end of the coil.

These effects are owing to a flow of electricity through the wire. You will remember what we learned (p. 211) about energy passing from one form into another, and about Chemical Affinity as one of the Forces (p. 204). Now the copper and zinc plates have a different chemical affinity for the acid solution in which they have been placed and which quickly begins to act upon one of them, and it is found that when the plates touch or are connected together by a wire, the energy of the chemical action is converted into that of *Electricity*, and a *continuous flow* or *current* of electricity is kept passing through the wire from one plate to the other. I said "connected together by a wire," but it would be more accurate to say "connected by continuous conductors," for wire is only one of the conductors of electricity, though perhaps the most convenient for ordinary use. You may have noticed already that when you laid the ends of the two wires on your tongue it was not necessary that they should touch each other for you to have the peculiar sensation given by the current of electricity; in this case, your moist tongue is the conductor which carries on the current from one wire to the other.

Such a jar as we have used, with its two plates and its liquid, is called a Voltaic or Galvanic *cell*, from the names of M. Volta and M. Galvani, the two famous discoverers of current Electricity. If we take a number of the cells and connect them together by short wires passing from the *copper* plate of each cell to the *zinc* of the next, we have what is called a *voltaic battery*.

There are several forms of voltaic cells, usually called after the name of the inventor, and devised one after another in order to get over some difficulty or drawback in the working of the previous cells. For instance, in the simple cell described above we should find that after a time the copper plate becomes entirely covered with bubbles of hydrogen gas, and this quite prevents the cell from giving any current. To cure this defect, various plans are adopted. One very good one, known from its inventor as Daniell's battery, is making the cell double,

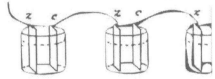

Simple form of voltaic battery.

so that the zinc should be still in sulphuric acid solution, but the copper plate be immersed in copper sulphate solution. This action of the hydrogen gas in adhering to the copper plate and stopping the current is called *Polarization*.

A current will run through any length of wire—miles of it—if the connection is complete; but if the bare wire is anywhere coiled in a loop, so that one part touches another, then the current instead of running all round the loop will take a shorter path across the points of contact, and return without all passing through the whole length of wire. To prevent this, which is called *short-circuiting*, wires used for electrical purposes are generally covered with cotton or silk or some other non-conductor, so that even if one part of the wire touches another no short-circuiting may take place.

If we take a bar of soft iron, and coil round it a large number of turns of covered wire, and then connect the ends of the wire with an electric battery so that a current passes through it, we shall find that the iron has become a magnet, and will attract small pieces of iron, but the moment that the wires are disconnected and the current ceases to flow, the iron loses its magnetic power. Such

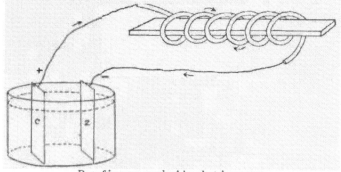

Bar of iron magnetized by electric current.

an arrangement is called an electro-magnet; it is only a magnet while a current is circulating through it.

We shall now, perhaps, be able to understand the principle of the electric Telegraph. Suppose we have an electro-magnet—that is, the iron and the coils round it—at one place, and then carry on the wires supported by telegraph poles to a town some miles distant. We may call the place where the magnet is Station A, and the place where the ends of the wire are Station B. At Station A let there be just above the electro-magnet a small piece of iron held by a spring, which keeps it about one quarter inch from the magnet. At Station B let there be a battery. If at B we connect the ends of the two wires, one to each end of the battery, a current will flow along the wire on the telegraph-poles to the magnet at A, and while it is running round the coils the iron bar will become a magnet, and attract down the small piece of iron suspended above it; but on disconnecting either wire at B the current stops passing, the bar ceases to be a magnet, and the small piece of iron is drawn up again by the spring. Now, by a system of signals arranged beforehand—signals depending on the longer or shorter time during which the iron is held down by the magnet—an alphabet may be made, and, therefore, words can be transmitted. This is the principle of one form of telegraph.

The earliest form of electric telegraph, and one still used in many places, is simply a magnetic needle surrounded by a coil of wire, like the arrangement described on p. 302, only for convenience the needle and coil are suspended vertically. Instead of long and short contacts from the distant operator, a code of signals—for every letter of the alphabet and for the figures o to 9, etc.—is arranged from the right and left movements of the needle, which occur when the current flows in one direction or the other.

From what was said on p. 302, you will understand how a delicately suspended magnetic needle inside a coil of wire can be used to detect the existence and to find the direction of an electrical current. Such an instrument is called a *galvanometer*, and by similar means we can also measure the strength of our electric current.

Another remarkable property of an electric current si

311

its power of chemical action; we shall refer to this in the next chapter. If you dip the ends of the wire from a voltaic battery into a slightly acid water you will see bubbles of gas coming off, due to the decomposition of the water. This electrical decomposition is called *electrolysis;* all liquids which are compounds (p. 313) and which transmit the current are thus decomposed into their two main constituents, one always going to the place where the current enters the liquid, and the other to the place where it leaves the liquid. Fasten a silver coin to each end of the wires from your battery and dip the coins, about an inch apart, into a solution of sulphate of copper; you will find the coin attached to the zinc end of the battery covered with a deposit of copper, and if you reverse the wires the copper will now be dissolved off the first and appear on the other coin. This is the principle of *electro-plating,* which is so largely used in the arts for depositing silver or gold.

Solids, owing to their cohesion, cannot thus be decomposed, or shaken asunder, but the molecules of the solid, if they resist the passage of the current, do get a shaking, which takes the form of heat. Bad conductors thus grow hot when an electric current flows through them, and if the resistance is great enough and the current strong enough the conducting wire becomes red-hot and even melts. This is the principle of the electric *glow-lamps,* now so familiar to us all; a thread of very resisting material (a filament of carbon) enclosed in a vacuum being employed. Another and more powerful method of electric lighting is the *arc-light.* Here the current is made to pass between two carbon pencils kept a slight distance apart, and in jumping across the gap great resistance is encountered and a very brilliant light is produced. The ordinary voltaic, or *primary*, batteries are not used to generate the current for the electric light, as they would prove both expensive and troublesome. Special machines are employed, driven by powerful steam-engines, and the principle on which they work is different from anything we have learnt so far.

If we have a coil of silk-covered wire, with the two ends of the bare wire twisted together, and quickly thrust one pole of a magnet into the coil, it will be found that the magnet starts a momentary current in the coil, which dies away again immediately. Pull the magnet pole out and another fleeting current runs through the wire, but in the opposite direction to the first. Observe that by this we find that just as an iron bar is made magnetic by being put inside a coil through which an electric current is passing, so on the other hand an electric current can be produced in a coil by moving it past the ends of a magnet. So electric currents produce magnetic power, and magnetism in turn produces electric currents.

It is not possible in this little book for beginners to explain the details of the arrangements by which, with the aid of rapidly revolving coils of wire and powerful magnets, machines are made for producing, on this principle, electric currents of enormous strength. All such machines are called magneto-electric machines, and another type dynamo-electric machines; some of them are of very great size, requiring engines of hundreds of horse-power to drive them.

CHAPTER XXI.

CHEMISTRY.

WE have already several times mentioned Chemical Affinity and Chemical Action, and must now turn attention more closely to Chemistry and see what kind of knowledge is classed under this name.

You know that if we allow milk to stand for a long time, especially if the weather is hot, it will turn sour. The milk will not be the same as it was when first taken from the cow; in other words, some change takes place in the milk. Again, when meat or eggs, or anything of that kind, "goes bad," as we say, what does that really mean? It means that these things pass through some alterations in which bad smelling substances take the place of good smelling substances. Now, what happens in all these cases? What was the original substance of the meat, eggs, milk, etc.? How came they to be changed? What are the new and bad substances? All these questions belong to the study of Chemistry.

Analysis of Water.—In the course of these lessons we have tried many experiments with our very useful friend water. We have seen it frozen into ice, boiled into vapour, and brought back from both these conditions to water again; we have dissolved things in it, and weighed it, and pumped it, and made it work for us. Now, let us take it once more, and try to find out what it is made of. For this purpose we will send through it a current of electricity.

Take a glass vessel with a hole in the bottom of it—a lamp-shade of this shape turned upside down will do

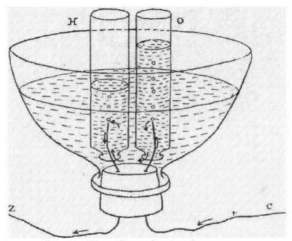

Decomposition of water by electric current.

admirably; stop up the smaller opening with a large cork, through which two wires have been passed, and pour water into the vessel, adding also a few drops of an acid to make the water a better conductor.* The outer

---

* The common but poisonous and corrosive acid called oil of vitriol—the chemical name for which is *sulphuric acid*—is best to use; but here, as in many other chemical experiments, great care must be taken not to spill any acid on your fingers or clothes, or a very bad burn will follow. The wires in the vessel should be of the metal *platinum*, for a reason that will be explained presently. The cork can, if necessary, be made water-tight by melting a little paraffin wax from a candle over it.

ends of the wire must now be connected with the wires of a voltaic battery (which must have at least two cells; these we have already described in the chapter on Electricity), and then an electric current will begin to flow round through the battery and the wires and the water. The moment it does so you will see, rising through the water from the surface and points of the wires, little bubbles of gas, which can be collected in the following way. Take a clean empty bottle (a narrow one with a small mouth will do best), fill it quite to the brim with water, and then, putting your thumb over the open top, turn it upside down with its mouth under the water in the basin, and bring it over one of the wire points. Another bottle must be filled and inverted in the same manner over the other wire, so that the gas-bubbles will rise into the bottles, turning out the water in them and taking its place. If the two bottles are placed over the wires at the same instant, we shall presently find that when one bottle is just filled up with gas the other is only half full. Now, where do these gases come from, and what are they? They come out of the water, which has been broken up by the electric current, and they are the actual substances of which the water was formed. If we were to mix these two gases together again and then bring a lighted match to them, they would "go off" with a big bang and a tiny drop of water would again be formed, the drop which was broken up, or, as we say, *decomposed* by the electricity. From this experiment, then, we learn the very important fact that water is made up of two gases, and that we can either obtain the gases from the water, or the water from the gases.

Now let us turn our attention to these two gases themselves. We will try some experiments on them and find out something about them.

Take the bottle which first filled with gas. Lift it out of the water, carefully keeping it still upside down, and wipe it lightly outside. Put a lighted match to its mouth, and notice that the gas will take fire and burn like the coal gas we use in our lamps, but the flame is very pale and blue. Now plunge the lighted match right into the bottle; it immediately goes out. This gas, then, will burn if set fire to, but it will not let a match burn in it. It is, as you can see, quite colourless and invisible, and it is also quite without taste or smell. The name given to it is *Hydrogen.* This is not the first time we have heard of Hydrogen ; in the chapter on Weight and Pressure (p. 226), it was spoken of as lighter than air. It is the lightest of all substances, its weight being less than one-fourteenth the weight of air, so that it is used to fill balloons which will rise and float in the air. In fact, it is so light that if you want to keep it in a jar even for a few minutes, the jar must be turned upside down to prevent it from all flying away upwards.

Now consider the gas in the other bottle, which is also colourless, invisible, and tasteless. Try and light it ; it will not take fire. Blow out the lighted match, or set fire to the end of a bit of string and blow it out, in either case so as to leave a red-hot tip ; plunge this into the gas, and the spark will burst out again into flame. The properties of this gas, then, are different from those of hydrogen, for it will not itself burn, but encourages other things to burn in it. It is also an old acquaintance ; we

have met with it several times under the name of *Oxygen.*
We found that when carbon is burning it is combining
with oxygen, that animals in breathing draw oxygen
into their lungs to purify the blood (p. 152), and that
plants in light are constantly giving it off (p. 195).

We have now split up water into the two gases of
which it is formed, and we find on examination that
these two gases are Hydrogen and Oxygen. This
process of splitting up a substance so as to divide it
into all the different kinds of matter of which it is
made is called *analyzing* the substance. We have
analyzed the water. The usual mode of analysis is by
chemical means, an example of which is given on p. 318.

Elements —Some substances have never been divided
or analyzed at all. Take, for instance, gold, silver,
oxygen, or hydrogen ;—do what we will to them, no one
has ever got anything out of them which is not gold,
silver, oxygen, or hydrogen. These are called simple
substances, or *Elements*, while those from which it is
possible to get more than one kind of matter, such as
water, marble, etc., are called *Compounds.*

We should hardly have guessed, by looking at it, that
water was not a simple substance; but there can be no
doubt of the fact when we have divided it into oxygen
and hydrogen, and reproduced it by putting these
together again.

Now, what other Elements are there found in the
world? Chemists have been at work for many years
examining and testing and analyzing substances, and
when they find one that they cannot by any art divide
further, they suppose that it is simple, and call it an

element; but, of course, more knowledge might some day show that they were mistaken about some of them. At present there are believed to be between seventy and eighty Elements or simple substances. Some of these have been discovered quite recently; in fact, during the printing of this book, the very air we breathe has been found to contain, besides its well-known gases, which form the bulk of the atmosphere, some *new elements* which have never been detected before. The number of Compounds is enormously large, quite past counting.

Combination: Mixture.—A chemical Compound, however, does not mean simply a mixture of two substances together, as we might mix sugar with salt. In such a mixture the particles of sugar and salt lie side by side unchanged, and if you taste it you can distinguish both the sugar and the salt at the same time. But when two substances are united, or chemically combined, as the proper phrase is, into a compound, they always produce something entirely *different* and with different properties from either of them singly. For instance, we have just found that hydrogen will readily burn, and oxygen readily allows things to burn in it; but when they are chemically combined they form water, which will neither burn nor allow anything to burn in it—indeed it extinguishes fire.

Take some very fine copper filings and mix them with some powdered sulphur. You can distinguish, at least with a microscope, the little particles of yellow sulphur and of red copper lying side by side; and if some of the mixture is thrown into a basin of water the sulphur will rise to the top, while the copper will sink to the bottom.

Now put some of the mixture into a small glass test-tube and heat it over a spirit-lamp; the sulphur will soon melt, the whole will get very red hot, and then quite a new substance will have been formed. Examine it; it does not look like either copper or sulphur; you cannot see small particles of either, but only a black mass, and if you throw it into water it sinks as a whole. This experiment shows the difference be-

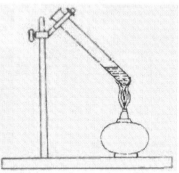

Sulphur and copper heated in a test-tube.

tween a chemical compound and a mixture. As long as the properties of the substances remain unchanged it remains a mere mixture; when a new substance with different properties is produced by their union, we know that chemical combination has taken place.

Several pretty experiments may be made to illustrate this point. Take some of a very strong solution of calcium chloride, and add to it very carefully some moderately strong sulphuric acid; both of these are colourless liquids, but, when put together, a white *solid* substance will be formed. The vessel into which they were poured may even be turned upside down, but nothing, or at most, a few drops of water, will run out. Here, then, is a solid produced from two liquids.

Again, get from the chemist some sal ammoniac, a white, crystalline, somewhat fibrous substance; powder some of it and mix it with some ordinary lime. Notice that neither of the substances has any appreciable smell. But, after mixing, and gently warming the mixture, smell it again—very cautiously, mind—and you will find that, owing to some portion of these two solids having chemically acted upon each other, a very strong smelling gas, ammonia, which is present in all smelling salts, has been set free.

So chemical changes may turn solids into liquids or gases, liquids into solids or gases, and also gases into solids or liquids.

Table of Elements.—The following table contains some of the most important among the substances recognized as Elements. You will see that they are arranged for convenience into two classes, metals and non-metals :—

Metals.

Iron.	Aluminium.
Lead.	Calcium.
Copper.	Magnesium.
Zinc.	Sodium.
Tin.	Potassium.
Gold.	Lithium.
Silver.	
Platinum.	
Mercury.	

Non-Metals.

Oxygen.	Silicon.
Hydrogen.	Phosphorus.
Nitrogen.	Sulphur.
Chlorine.	Bromine.
Carbon.	Iodine.

Metals.—Of these *Metals* those in the first column have long been known, and are among the most useful to man in the various arts and manufactures. The metals in the second column have all been discovered during the present century; they have many properties in common, they are never found pure or "native" in the earth, as they combine so easily with oxygen, and all of them when combined with oxygen, silicon, etc., form an important part of the solid rocks and soils of the earth, Aluminium being the principal ingredient in clay, Calcium in chalk, etc. Sand and quartz, however, contain only the non-metal Silicon combined with oxygen.

The Metals, as a rule, are heavy, that is, heavier than water, and will therefore sink to the bottom if put into a basin of water; but Lithium, Sodium, and Potassium are very light, and will float on the surface. Moreover, they decompose water, dividing it into oxygen and hydrogen as the electric current did; and when water is decomposed in this manner the hydrogen which comes off usually takes fire, while the oxygen combines with the metal, and forms a compound which will dissolve in the water and make it feel soft and soapy. These three curious metals are so soft that they are easily cut with a knife, and Lithium may even be spread like butter.

Lithium is the lightest solid known. The heaviest of the well-known metals is Platinum, which is more than 21 times as heavy as water. Gold is 19 times, Mercury, or quicksilver, 13½, Lead 11, Silver 10, Copper 8, Iron 7½ times as heavy as water, while Lithium is only half the weight of water.

Non-metals..—Of the above non-metallic Elements four are gases—oxygen, hydrogen, nitrogen, and chlorine. Of the first two we have already learnt something. The last, chlorine, is a yellowish-green gas, has a very bad smell, and will kill any one who breathes it ; even a little of it produces violent coughing. We will leave it alone at this early stage of chemical study; but nitrogen is very important to us, and must be examined more closely. In fact, the atmosphere, the air we breathe, is made up of oxygen and nitrogen; both gases retaining their own properties, so that it is not a chemical combination, but a mixture.

We can prepare nitrogen by taking away the oxygen from the air; and a very simple means of doing this, and one you can try for yourself, is as follows.* Get some clean iron filings or turnings and tie them up in a little muslin bag ; by means of a stiff piece of wire or a stick support the bag of filings at the top of a tall glass jar, standing upside down in a basin of water, as shown in the picture. The jar is of course full of air, though called "empty." If you have not got a glass jar, a large wide-mouthed bottle will do, or a lamp chimney tightly corked at one end. Just before making the experiment, moisten well the little bag of filings (it is best to dip them into a solution of sal-ammoniac), now put the jar or bottle over them, and leave the whole standing in the position shown in the picture till the following day. When some 24 hours have elapsed you

* The usual and the quickest way of preparing Nitrogen is to burn a piece of phosphorus inside a jar of air. The phosphorus combines with the oxygen, forming white fumes of an oxide of phosphorus which rapidly dissolve in water. If you try this experiment, get your teacher's help, for phosphorus is a dangerous substance for you to handle, and requires to be kept and cut under water, and not touched with the fingers, as it catches fire so very easily and produces a horrible burn. The experiment with the iron filings is quite easy and harmless.

will notice the water has risen in the jar; in a cylindrical jar 10 inches high you will find just 2 inches of water have entered the jar. What is the reason for this? The explanation is that the iron filings, or some part of them, have combined with the oxygen of the air inside the jar, and are thereby changed into an oxide of iron, a black rust, and the gas now left in the jar is no longer air, but Nitrogen. The water has risen to take the place of the oxygen which has disappeared, and as the water

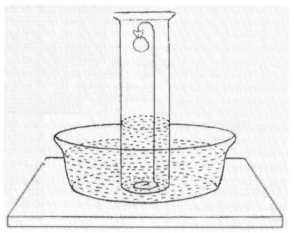

Preparation of nitrogen from atmospheric air.

now occupies one-fifth of the whole volume of the jar, this shows the important fact that only one-fifth of the air consisted of oxygen. The remaining four-fifths is Nitrogen.

Now slip a card, large enough to cover the mouth of the jar, beneath the water; holding the jar with one hand, and pressing the card against the mouth of the jar with the other, deftly turn the jar upside down, and let it stand on the table. Light a match or paper spill, and, removing the card, dip the light into the jar; it instantly goes out. The gas neither takes fire nor allows the match to burn in it. Make this experiment, and repeat it, in a quiet deliberate manner. Now see if you can taste or smell the gas. Breathe a mouthful; it has no taste nor smell, and certainly it has no colour; like the air, it is quite invisible. Nitrogen gas seems, indeed, to be characterized by the number of things it *cannot* do, and to be so uninteresting a substance that it is rather surprising to find that it forms so large a proportion of all the air. Its great use in the air seems to be to dilute the oxygen, and prevent it from combining with substances as rapidly and violently as it would do if there were not something to moderate its action.

But although Nitrogen itself is not very interesting, the case is quite different with its compounds, which have very conspicuous properties. Many, for instance, of the most brilliantly coloured dyes are compounds containing nitrogen; so are gunpowder, dynamite, nitroglycerine, and other explosives; and among yet other nitrogen compounds are some of the most violent poisons known, such as morphia, strychnine, and prussic acid. The strong-smelling ammonia gas is made of nitrogen and hydrogen; it very easily dissolves in water, and a solution thus formed is the ordinary liquid ammonia of the chemist's shop. Nitric acid, a compound of nitrogen, hydrogen, and oxygen, is one of the strongest acids known, very poisonous, and violently corrosive. If dropped on to copper, silver, or zinc, it eats into them very quickly; and if it gets on to the skin it burns that too, and may leave a bad sore, so that it must be handled with the greatest care.

Besides the four gases, the only other non-metallic Elements are Carbon, Bromine, Iodine, Fluorine, Sulphur, Phosphorus, Silicon, Boron, Selenium, and Tellurium, so that they are but few in comparison with the metals.

Carbon is a solid substance, which forms, when combined with oxygen and hydrogen, almost all vegetable matter; plants contain also combined nitrogen, and the flesh of animals consists of the same four elements. Coal, which is the remains of old plants pressed together and buried in the earth, is therefore very rich in carbon, and coke or charcoal is almost pure carbon. So is the mineral named graphite, or the so-called black lead in our pencils—which really has nothing to do with lead; and another form of pure carbon is the diamond, the most beautiful and precious of jewels.

The carbon compounds burn readily, and we use them to procure heat and light. Coals and coal-gas, wood, paper, alcohol, or spirits—which are all vegetable products—tallow, and other animal and vegetable fatty and oily substances, and also the so-called mineral oils, paraffin, naphtha, etc., are all compounds of carbon.

It is a curious thing that the same pure Element should be able to exist in various forms, and the account of how it happens belongs to more advanced chemistry; but carbon is not the only element which behaves thus, sulphur and phosphorus and some other things being also found pure in different conditions. The name given to this peculiarity is *Allotropy :* charcoal, graphite, and diamond, are called allotropic forms of carbon.

Ordinary *Phosphorus* is a yellow waxy substance, poisonous, and, as we have seen, so inflammable that it must always be kept under water. The other, or allotropic form of phosphorus, is a dark red powder, not at all so inflammable, nor so poisonous—very different, therefore, from the first variety. The tips of ordinary matches are covered with a phosphorus mixture, which easily ignites by the heat produced in striking the match: with safety matches, however, which strike only on the box, no phosphorus is placed on the match head, but the brown preparation on the box chiefly consists of the red phosphorus.*

Phosphorus is obtained from bones, the earthy part of which is a compound of phosphorus, oxygen, and calcium, called phosphate of calcium.

Most of us are familiar with the ordinary form of *Sulphur*, or Brimstone, a yellow, solid, light substance. If a little is placed in a test-tube and heated gently, it will first melt and form an amber-yellow liquid; when further heated the liquid becomes darker in colour and thicker, so thick, indeed, that the tube may be turned upside down and nothing will run out. On still further heating, it gets almost black and thinner, and finally it boils. If, just as it begins to boil, it is poured out in a thin stream into some cold water, it will instantly

---

* A terrible disease of the jaws is liable to be produced in the men and women who are engaged in making the poisonous common match, so that in several civilized countries, but, alas! not in England, their manufacture is prohibited. The safety matches are harmless to the maker.

become solid, but will form a yellow stringy mass just like india-rubber. This is called plastic sulphur, and is an allotropic form of sulphur.

Bromine is a dark red liquid, with a very bad smell, the only element except Mercury or Quicksilver which is liquid at ordinary temperatures; and *Iodine* is a bluish black solid, smelling something like Bromine, only not so bad. The elements, Chlorine, Bromine, and Iodine, are very like one another in their properties.

We have spoken of only a few out of the long list of elements, some of which are very rare, and many others, though not without importance, are not found in very large quantities, so that, in fact, the greater part of the earth and the things upon it (including the water and the air) are principally composed of some ten or twelve elements, combined together in many different forms and proportions.

Of these Oxygen is by far the most important. Mixed with nitrogen, it forms the air; combined with hydrogen, it produces all the water in the world; it is an essential part of all vegetable and of all animal matter, and in its various combinations it forms fully half the material of the solid crust of the earth.

Oxidation.—Oxygen can combine with almost all the elements, and a special name is given to the substances composed of an element combined only with oxygen; they are called *oxides*.

When a piece of bright iron is left out in the air, it will soon rust, as we say, especially if the weather is damp, and the rust is merely a compound of the iron with some of the oxygen in the air; there is a black and a red oxide of iron; the latter, the red rust, is called in Chemistry Ferric Oxide.

Get some magnesium ribbon at the chemist's shop, cut off a piece about six inches long, hold it by a pair of pincers over a plate, and set fire to it; it will burn with a very brilliant white light, and when it is burnt out, instead of the bright shining metal we shall find on the plate only some white powder. This is a compound of oxygen and magnesium called Oxide of Magnesium, or Magnesia.

If we put a small piece of the soft metal, Sodium, into a spoon and hold it over a flame, it will melt, take fire, and burn with a yellow light, and a white powder will be formed which is Oxide of Sodium, or Soda. We may notice that when the piece of sodium was cut off for this experiment there was barely time to see the bright newly-cut surface of the metal before a thin layer of the oxide began to form over it, and make it dull. This is what is meant by saying that a metal tarnishes. Silver does not tarnish easily, that is, it does not readily form oxides by simple exposure to the air; gold and platinum do not tarnish, or oxidize, at all in this way, only with difficulty can they be made to unite with oxygen. Hence their great value for many chemical as well as ornamental purposes; and so we see that elements differ from each other in the ease and rapidity with which they combine with oxygen.*

* If iron wires had been used in the experiment of decomposing water by an electric current (p. 310) the oxygen set free would immediately have combined with the iron; hence it is necessary to use platinum wires, because platinum does not combine directly with oxygen.

The examples just given are all of them oxides of *metals*, but oxygen combines also with the *non-metals*, and the oxides so formed are very interesting in their properties.

We saw one of them in the experiment with burning phosphorus : the white fumes with which the glass was filled by the burning consisted, as we said (see footnote on p. 318), of an oxide of phosphorus. In our experiment this was all presently dissolved in the water, but if the oxide is wanted for separate examination the experiment may be repeated on a dry plate, when the flames will settle as a coating over the inside of the glass. After the phosphorus has burnt out, the glass can be lifted up, and the white powder scraped off into a little heap on the plate. A drop of water let fall into this heap combines so rapidly with the oxide that it will hiss as it would when falling on red hot iron.

Hence, after burning phosphorus over water, the white fumes that dissolved in the water and disappeared, formed a fresh compound made of oxide of phosphorus and water, which is called Phosphoric *Acid*.

Acids.—Acids have a peculiar sharp, sour taste, which we can only describe as an acid taste, but with which we are all very familiar in such compounds as lemon juice, vinegar, unripe apples, etc. It would not be safe to try all acids by tasting them, but a good test is found in a substance called Litmus, which changes colour when touched by an acid, so that a purple or blue solution of litmus, or a piece of paper coloured blue by litmus, will turn red if any acid is poured on it. If pieces of zinc or iron are put into an acid solution, hydrogen gas is always given off, the result of decomposition of the water. This action, however, though characteristic of acids, is not equally strong in all, and phosphoric acid does not show it at all so well as sulphuric. In fact, the best method of obtaining hydrogen for experiments is not by passing an electric current through water, but by pouring sulphuric acid largely diluted with water on small pieces of zinc, when bubbles of hydrogen begin rapidly to rise through the water : the zinc robs the water of its oxygen, uniting with it, and then with the sulphuric acid, forming a substance called sulphate of zinc, which is quickly dissolved by the water.

Now, notice this. Oxide of phosphorus combined with water produces *Phosphoric acid ;* oxide of sulphur with water produces a strong, burning acid, called *Sulphuric ;* oxide of nitrogen with water produces *Nitric acid.* But phosphorus, sulphur, and nitrogen are non-metallic elements, and so we find that the oxides of *non-metals* with water produce the compounds called *acids.*

Alkalies.—Compare the oxides of the metals with these acids. Take some of the white powdered Soda, or oxide of sodium, which was produced by burning sodium, and put it into a vessel of water; it dissolves easily in the water, though not quite as quickly as the oxide of phosphorus did, and forms a solution of soda. Taste this solution : it has a queer taste of its own, rather soapy, but certainly not acid. Try its effect on litmus; it turns the red litmus blue, seeming to have just opposite properties from those of acids. It belongs to a class of substances we call *alkalies.*

Now mix it with an acid solution. To a strong

solution of soda add very carefully some strong nitric acid; you will notice that when these two cold liquids are put together they become quite hot. Heat is always given out when substances combine chemically. If we test the combination with litmus we find that there is no change of colour, provided we have put exactly the right quantity of acid—this the litmus paper tells us; the acid will not turn it red, and the alkali will not turn it blue; they have combined and neutralized each other's effects.

But as the solution gradually cools some fine white crystals will be formed in it. We want to observe these particularly, so we will carefully pour off the liquid and examine them. The substance does not taste at all like the acid or the alkali—it is quite neutral; it is called in chemistry *a salt*, and the name of this particular salt prepared by the action of nitric acid on soda is *sodium nitrate* or *Chili saltpetre.*

Salts.—There are many different salts. Ordinary "bluestone" is a salt formed by the action of sulphuric acid on copper oxide; the "salt" we eat is a salt which may be formed by the action of hydrochloric acid on soda, but most of that which is used in commerce is not prepared by chemists, but is obtained from the earth.

Bases.—You see we have formed a *salt* by the union of an acid with an alkali. But many of the oxides of the metals are insoluble in water, and yet when dissolved by acids they produce salts. Another name is therefore given to all the substances which neutralize acids; they are called *Bases*, and if soluble in water are called ALKALIES.

332

Let us see what we have learned so far. First we tried to understand what is meant by an element, or simple substance, and saw that the elements were divided into metals and non-metals: then went on to consider the oxides, or compounds formed by the different elements with oxygen—the oxides of the *non-metals* which, with the addition of water, produce *acids*, and the oxides of the *metals* which, if acted upon by water at all, give strong *alkaline* solutions, but are, in any case, called *Bases ;* and finally saw that the Bases, or metallic oxides, in combination with acids, produce *salts*.

Combustion.—We had occasion earlier in this book to talk of substances combining with oxygen, and I want you now to go back to Chapter XII., p. 210, and look again at what was said there. We were talking of Forces and Energy, and found (1) that one of the forces which can give rise to energetic action is the Chemical affinity, or tendency to combine with each other, that exists in different substances, and (2) that until it is satisfied by bringing about the desired combination it is in the condition of *potential* or stored up energy, waiting to act; but when it is expended or paid out in doing the active work of combining, it goes off into some other kind of energy, heat, light, and electricity being the most usual forms.

Let us see what happens when a fire is lighted. The fuel that is piled in the grate is full of latent or hidden power, for it is always waiting for a chance of combining with oxygen; and there is plenty of oxygen in the air all round, only while the fuel is cold it is not in a favourable condition for combination. When, however, we bring a lighted match into contact with the paper, it is warmed enough to rush vigorously into combination with the oxygen, and the energy is changed into so much heat that the faggot is heated and the wood begins combining with oxygen, or "burning." This combination again sets free much more energy, as heat, which, in its turn, starts the combining of the coals, and so it goes on till everything combustible within reach, that is, everything which heat can readily enable to combine with oxygen, is combined or burnt up. The paper, wood, and coals all contain carbon, and the new substance produced by the combination, or burning, is an oxide of carbon, commonly called carbonic acid gas, which we will presently examine.

In the great heat caused by the energetic rush of combining, some parts of the solid carbon are raised to such a temperature that they become luminous, or give out light. The glowing red hot coal is a luminous solid, the flame that dances over it is luminous gas, made more brilliant by containing tiny solid particles of white hot carbon.

Just the same thing happens when a candle burns. It is hot because the wax is combining energetically with oxygen, luminous from the high temperature of the gas that is being formed, and brilliant from the white hot particles of carbon floating in the gas.

The coal and the candle do not consist of carbon only; there is hydrogen combined with the carbon, and the hydrogen being set free in the burning combines also with oxygen and forms water vapour. If we hold a cold, bright glass tumbler over a candle flame, it immediately becomes misty with fine steam, because the cold glass condenses some of the vapour into water. All the variety of paraffins and other oils that we burn in our lamps are also combinations of hydrogen and carbon in different proportions; and you have very likely noticed that when a lamp is first lighted, and the cold glass chimney placed over the flame, it turns misty, like the tumbler, with the steam, but in a minute or so the glass grows too hot to condense the water and so becomes clear again.

None of the matter is lost or destroyed by burning; it only changes its combinations and exists in other forms; and if when a candle is burned we could collect and weigh all the carbonic acid gas and water vapour formed by its burning, we should find that their weight was exactly that of the candle added to the weight of oxygen with which it had combined.

You see that we have several times here spoken of "combining *or* burning," and I want you to notice that "burning" is only another name for combining, usually with oxygen. Some substances oxidize with a swift violent energy which produces great heat and light, and some oxidize quietly and slowly, though more or less heat is almost always given out in the act of combining. In common talk we only give the name of "burning" to the rapid, hot, bright oxidizing; but chemists will tell you that all oxidizing, even the quietest, is really "burning," or *combustion*, which is the proper word for it.

Respiration.—Do you remember when we first had occasion to learn the name of carbonic acid gas? How we saw that in *breathing* we draw fresh, clean air into

335

our lungs, and there the blood, seizing upon the oxygen in the air, carries it along in its ceaseless travels (p. 151).' What the oxygen does in the blood is to combine with the particles of waste carbon in all parts of the body, forming carbonic acid gas, which the blood carries back to the lungs to be breathed out and so got rid of. We know enough now to call the process by its proper name of combustion or burning, and can see that oxidizing the waste carbon is simply burning it up, just as the carbon in the coals and the candle are burnt, and the heat given out in the combination is what keeps our blood warm and so warms our bodies. The heat is not so great as to cause flame and fire, because the combustion which is always going on is spread about all through the blood, and not concentrated into one spot. So all the warmth of living animals is produced by chemical combination.

The effects of rapid oxidation are sometimes very sudden and startling indeed. Gunpowder is a *mixture* of sulphur, charcoal or carbon, and nitre or potassium nitrate. These substances are powdered, mixed together carefully, and then the mixture, after having been wetted, is squeezed together by machinery with very great pressure, a process which turns the moist powder into a hard slate-like material which can afterwards be broken up into pieces of any desired size. When gunpowder is set on fire, the large quantity of oxygen in the nitre combines with the carbon and sulphur, and forms in an instant a very large volume of gas which, expanding suddenly, has power enough to hurl aside anything that has confined it. The explosion of gunpowder is, therefore, nothing more than the very quick burning of carbon and sulphur.

Carbonic Acid Gas.—Now let us examine this carbonic acid gas, which chemists call carbon dioxide.

We can get it, as we know, by burning carbon in the air, but another and a convenient way of preparing it is by putting some pieces of chalk or marble into a wide-mouthed jar, such as a pickle bottle, and pouring gently upon them some hydrochloric acid and water; the acid must be carefully handled, as it is poisonous. There will instantly be a fizzing, and the jar will soon be full of the gas, with a little liquid at the bottom. Marble and chalk are different forms of what is called calcium carbonate, a compound of the metal calcium with carbon and oxygen, while hydrochloric acid is a compound of hydrogen and chlorine dissolved in water. When they are put together the carbon and some of the oxygen form carbonic acid gas, the rest of the oxygen and the hydrogen become water, and the calcium and chlorine combine into calcium chloride, which is left dissolved in the water at the bottom of the jar.

If we test the carbonic acid gas as we did the oxygen and the hydrogen, we shall find that it has neither colour nor smell, that it will neither burn nor allow anything else to burn in it. Nor indeed can anything live in it; a living creature that gets into it will soon die.

When we wanted to keep hydrogen in a jar, we found it necessary to turn the jar upside down, because the light hydrogen would all rise into the air and fly away, but this gas does not show any sign of doing so.

Put a second jar, empty and clean, beside the one

we are using, light a match or a taper, and dip it first into the empty jar, where it burns easily, and then into the gas, where it instantly goes out. Now take the jar of gas and, holding it over the mouth of the empty jar, pour carefully from one to the other; but take care not to let any of the liquid at the bottom pass out. If, after this, you test again with the lighted match, you will find that it is now extinguished in the second jar— showing it contains carbonic acid gas; the gas has been literally poured from one jar to the other, proving that it is a very heavy gas.

We have produced some carbonic acid gas from chalk; now let us go the opposite way to work, and see if we can get chalk again by means of the gas. For this purpose we will put some pieces of ordinary quick-lime, which is an *oxide* of calcium, into a large bottle, nearly fill it up with water, shake, and then allow it to stand for two or three days; after which time the solid lime will have sunk to the bottom, leaving above a clear liquid, which is lime water—that is, water with some of the lime dissolved in it. We can pour off the lime water into an empty jar, leaving the sediment behind. If this is done steadily and without shaking, the lime water remains quite clear. Now, however, let us pour a little of it into a jar containing carbonic acid gas, and we find that the lime water instantly turns milky. The reason of this is that the carbonic acid combines with the lime in the water and forms chalk (calcium carbonate). Chalk will not dissolve in water like lime, and the tiny particles of solid chalk suspended in the water cause the milky appearance.

338

We can make the lime water look milky just as well by simply blowing into it some of the carbonic acid gas from our lungs. Let us take some of the clear lime water in a test-tube, and blow gently into it through a piece of glass tube until it becomes milky ; do not stop then, however, but continue blowing, and in a few minutes all the milkiness disappears and the liquid is clear again. What is the reason of this ? It is that chalk, though it will not dissolve in pure water, *does* dissolve in a solution of carbonic acid. Our first breaths of carbonic acid only converted the lime into chalk ; but when this change was complete the chalk would not take up any more of the carbonic acid, which went on passing into the water, and soon brought it to a condition in which it could dissolve the chalk. Our clear water now is, therefore, not the same as the clear water we began with. That contained dissolved lime, but this contains chalk dissolved in carbonic acid.

If we now suspend the test-tube over a spirit lamp and make the water boil, the excess of carbonic acid gas is driven off in the boiling, and the chalk, no longer having anything which can dissolve it, reappears in milkiness.

Hard and Soft Waters.—We will test our clear waters in yet another way. Here is a tub that has been left out in the garden, and is half full of rain water. If we put some into a basin and wash our hands in it they are quickly clean, for the soap lathers up directly ; indeed, if we use a good deal and beat up the water for a minute it becomes a mass of soap bubbles.

Now try washing in the water which contains dissolved

chalk, and we shall find the difference; at first we cannot get a lather at all, but presently there will come little flakes like curd, and after a little perseverance the soap will lather. This is what we call a *hard water*, while the rain gives us a *soft water*. *Pure* water is always soft. But if our water is hard through the presence of chalk dissolved in carbonic acid, we know now how to soften it. By boiling the water (as we saw just now in the test-tube), the excess of carbonic acid is driven off, and the chalk, no longer dissolved, falls down as a white mud, forming a white coating, or furr, on the inside of the kettle or boiler, but leaving the water fairly pure and therefore *soft.*

This kind of hardness which may be removed by boiling is called *temporary hardness.* But if instead of chalk the water contains dissolved gypsum (sulphate of lime), or common salt, like sea water, this cannot be removed by boiling, but gives rise to what we call *permanent hardness.*

Diffusion of Gases.—Since all burning forms carbonic acid, and all animals breathe out this poisonous gas, you may perhaps wonder where it all goes to, and why, as it is so heavy, it does not all lie about on the surface of the earth, and kill all animals, and put out all fires.

Well, there are two reasons why this is not the case. First, it is not possible to keep a gas by itself at all, unless it is tightly shut up in some vessel; if any other gas can reach it, no matter how much difference in weight there is between them, they will gradually mix together. Heavy gases will go upwards, if necessary, and light ones will come down, each diffusing itself

340

through as large a space as possible. It is true that we were able to pour carbonic acid gas from one vessel to another, and also that when hydrogen was produced by the action of sulphuric acid on zinc, we caught some of it in a jar turned upside down. But in both these cases, we have to experiment quickly with the gases before they are lost by diffusion in the air, which will soon occur unless they are kept bottled up. So we see that even the heavy carbonic acid gas, if only it has free access to the air, gradually diffuses itself into the whole atmosphere. Where, however, it is confined with little opportunity of mixing with air, it becomes a real danger. In coal mines it forms the terrible *choke-damp*, and it frequently occurs at the bottom of old wells; even rooms where several persons have been breathing become unwholesome with carbonic acid, unless pure air is freely admitted so as to mix with it. Winds and convection currents (p. 253) are energetic helpers in this mingling of the gases of the atmosphere.

But there is a second and very important process going on, by which carbonic acid is not merely dissipated but actually destroyed. This is the action of plants, which when exposed to sunlight absorb carbonic acid from the air, and separating the oxygen and carbon, send back the oxygen into the air while they feed themselves upon the carbon (see p. 195). Thus we find that while men and other animals absorb oxygen and breathe out carbonic acid, plants and trees absorb the carbonic acid and give out oxygen, so purifying the air and making it fit to be breathed again by men.

Summary.—You will observe that almost the whole

of what we have learnt in this chapter is about Oxygen and its combinations—how with metals it forms *bases*, and with non-metals *acids*, and how bases and acids combine into *salts ;* how the process of combining with oxygen is the same thing as *combustion ;* and finally, we considered the slow form of combustion called *respiration* or breathing, and the carbonic acid gas which is produced by respiration.

But, as there is not time or space to go further into Chemistry here, the whole subject of combinations between the other elements, as well as the question of the fixed numerical proportions in which elements combine, and many other important laws of chemistry, must all be left for future study.

In the last ten chapters we have made a beginning of the study of the vast and various forms of Energy among which we live and move. And we see that, though we are not able to create or destroy these natural Forces, yet we learn by experience first to adapt ourselves so as to live at peace with them, and then by degrees to guide and transform and use them for our own purposes. Yet men's knowledge is still very limited, and it is open to every thoughtful and accurate observer of nature to do something to increase the sum of human knowledge and power.

THE

CHEMISTRY OF CREATION:

BEING

A Sketch of the Chemical Phenomena

OF

THE EARTH, THE AIR, THE OCEAN.

By ROBERT ELLIS, F.L.S.

M.R.C.S. ETC.

THE ALCHEMIST.

INTRODUCTION.

WE must look through a long vista of ages if we would discover, buried in the obscurity of time, the origin of what is now called the science of Chemistry. We know little about the period when the few facts which formed its first beginning were gathered together, but it appears probable that Egypt was the country where this took place. At least, it is nearly certain that Chemistry meant originally Egyptian or secret knowledge, as it was afterwards called the secret or the black art. Plutarch states that the old name for Egypt was Chemia, and that the name was given to it on account of the blackness of the soil.

The science of Alchemy had its origin in Egypt, and its object was the production by artificial means of the noble metals. In the fourth century we meet with the word Chemia as the name of the art of making gold and silver.

The Arabians, who, in the year 640 overran Egypt, there learnt their chemistry, and they prefixed the Arabic article *al* to *chemia,* and thus made Alchemy a familiar term for the transmutation of the base into the noble metals. A celebrated physician, of the name of Geber, paid great attention to Alchemy, and discovered some most important substances. The language used by him was often very high-flown, and, indeed, intended to disguise the processes employed from the unlearned : the name gibberish has probably its origin from Geber. Many chemical operations are described by Geber, such as distillation, filtration, sublimation, and crystallization; and by these he prepared new substances.

The accompanying hieroglyphics assure us of the fact that the Egyptians knew how to blow glass in the same manner as we do; and that thus have been formed useful chemical vessels for the early professors of this art. So far had the glass-workers of Egypt advanced in their art, that even coffins were sometimes made of glass.

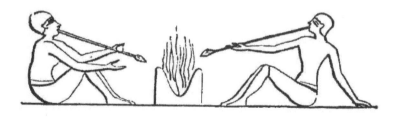

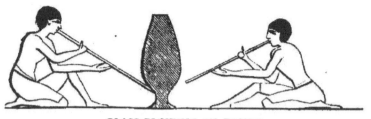

GLASS-BLOWING IN EGYPT.

From the Arabians the knowledge of alchemy spread over Western Europe, and in the thirteenth century we find alchemists of the Arabian school in all the chief countries. Whilst Albertus Magnus flourished in Germany, Roger Bacon was well known in England. He was tried at Oxford for sorcery, and, to refute the charge, wrote the famous treatise in which he shows that appearances then attributed to supernatural agencies were due to common and natural causes. All the alchemists agreed that the transmutation of metals was possible with the aid of the philosopher's stone. One of them tells us :—" I had long doubted whether gold could be made from quicksilver. One who wished to convince me of my error sent me a drachm of a certain powder, of a red colour, having a peculiar odour, with which I was to make the experiment. To avoid the possibility of fraud, I purchased the requisite vessels and materials from an ordinary warehouse. I put the mercury into the vessel, and cast the powder into it. A strong heat was then applied, and immediately the whole mass was transmuted into ten drachms of the finest gold!" We are even told in history that a celebrated philosopher, in the presence of King Edward VI., by means of a certain powder converted a mass of iron into gold, which was afterwards coined into rose nobles. The powder of the true philosopher's stone was so powerful, it was said, that a few grains of it would turn twenty tons of lead into gold. If only the Chancellor of the Exchequer possessed some of it!

We need scarcely say that the philosopher's stone was never discovered, and the art of making gold usually ended in reducing its professor to rags. Its vanity and certain results are well told in the following lines of Spenser :—

" To lose good days, that might be better spent ;
    To waste long nights in pensive discontent ;
    To speed to-day, to be put back to-morrow ;
    To feed on hope, to pine with fear and sorrow ;
    To fret their souls with crosses and with cares ;
    To eat their hearts, with comfortless despairs ;
    Unhappy wights ! born to disastrous end,
    That do their lives in tedious tendance spend."

It was a striking example of that unquenchable hope which will hope against hope, that the idea of an elixir conferring immortality could ever have long occupied the attention of philosophers. The origin of this remarkable error admits of being traced, like that of so many others. "And now, lest he put forth his hand, and take also of the tree of life and live for ever," may have induced men to believe that something eaten or drunk would confer immortality. Certainly this hope continues to this day, as may be witnessed in the popularity of many patent medicines which answer to the elixir of the alchemist.

The last of the three delusions was the alcahest, or universal solvent. It may be considered also the most harmless. Properly speaking, it was simply a foolish phantasy of Chemistry. The idea was that some fluid might be produced which would instantly dissolve all substances exposed to its influence ; and it seems to have had a long existence as a fanciful speculation, rather than as a subject of arduous experiment and tedious research.

In the sixteenth century Paracelsus effected the union between Chemistry and Medicine, and this continued up to the end of the seventeenth. The Aristotelian doctrine of air, earth, fire, and water held its ground until the seventeenth century, when it was disputed by Van Helmont. He denied that fire had any material existence, or that earth can be considered as an element; but even he admitted the elementary nature of air and water. It was he who made the important discovery that if a metal be dissolved in an acid, it is not destroyed, as was the general belief, but can again be obtained as a metal. It was about the middle of the seventeenth century that philosophers laid the foundation of modern Chemistry. The principles laid down in the celebrated work called *Novum Organum*, of the illustrious

Chancellor Bacon, proved most beneficial to the development of true knowledge ; and as chemical philosophy was among the earliest to benefit by these principles, so it soonest began to expand and gather strength. The origin and further progress of the science has been happily compared to Milton's fine description of the erection of Pandemonium :—

> . ·. . . . " Soon had his crew
> ` Opened into the hill a spacious wound
> And digged out ribs of gold. ; . .
> Anon out of the earth, a fabric huge
> Rose like an exhalation. . . . .
> Built like a temple."

The progress of knowledge received a powerful impulse in the foundation of the Royal Society in 1662, and of the Academy of Sciences at Paris in 1666.

Robert Boyle was the first to make the distinction between elements and compound bodies. He maintained that all bodies are to be considered as elements which are themselves incapable of further separation, but which can be obtained from a compound body, and out of which the compound can be again made up. He upheld the view that the simple advance of knowledge was the highest aim of all research. "My kingdom is not of this world. I trust that I have got hold of my pitcher by the right handle, the true method of treating this study. For the pseudo (or false) chemist seeks gold; but the true philosopher, science, which is more precious than any gold." It was in the same spirit that a just reproof was given by D'Alembert to an ambitious young man ; and, as it deserves remembering, we venture to record it. ".Science," said he, "must be loved for its own sake, and not for any advantage to be derived from it. No other principle will enable a man to make progress in the sciences."

It would be impossible, in such an introduction, to do justice to the labours of Stahl, or to do more than allude to the phlogistic theory or to phlogiston, the combustible principle. Bodies like quicklime, unaltered by fire, were considered to have already undergone combustion; if phlogiston were added, the metal would result. When metals are calcined, an earthy residue or calx remains ; metals are therefore compounds containing a principle (phlogiston) which causes their combustion. Sulphur and phosphorus are compounds containing a principle which causes their combustion. Stahl did not take account of the fact, proved by Boyle, that metals increase in weight when they are burnt. Nevertheless, Stahl prepared the way of the modern Chemistry.

It was on August 1st, 1774, that Joseph Priestley discovered oxygen, during his stay at Lansdowne House, Berkeley Square. He obtained it in heating red precipitate by the aid of the sun's rays concentrated with a burning-glass ; and he called it dephlogisticated air, because so free from phlogiston. In 1781, Cavendish proved the composition of water, and showed that dry air was composed of 20.8 measures of dephlogisticated and 79.2 measures of phlogisticated air.

In the same year of the discovery of oxygen, Scheele discovered chlorine, which he called dephlogisticated muriatic acid.

Interesting it would be, but altogether out of place, to follow in consecutive order the further progress of the science; and we shall, therefore, hasten to its close. Lavoisier soon put an end to the phlogistic theory. Dalton gave us the atomic theory, proving that the atoms of the different elements are not of the same weight, and that the relative atomic weights of the elements are the proportions by weight in which the elements combine. Gay-Lussac discovered the law which regulates the combination of gases by volume. Sir Humphry Davy proved the compound nature of the alkalies. Berzelius proved that organic bodies obeyed the same laws as inorganic, and Wöhler showed the artificial preparation of a substance supposed to be peculiar to animal life. In the hands of the illustrious Liebig yet further progress was made, and this progress still continues.

In considering the present aspect and relations of chemistry, we are struck with the extent of its influence, and with the importance of the position it occupies. Advancing years are continually extending the one and augmenting the other. Every branch of the arts now experiences its salutary reign. While it has contributed much to the growth of other sciences, by no means directly or in the abstract related to it, it has also added a variety of substances to our present list of domestic comforts and conveniences. While it has tinged the purple and bleached the fine linen of the great, it has endowed with equal snowiness, and an equally durable though more homely lustre, the calico and coarsest fabric of the poor.

Nor has it been less valuable in adding to our remedies for the sick. For medicine, in fact, it will in future time do more, and this by reason of its intimate connection with that art, than for any other department of science. In many instances, chemistry detects the disease and points with much significance to the appropriate remedy. It analyzes the processes constantly in operation in the mysterious laboratory of the human frame, and indicates with precision many of the changes which matter undergoes in the performance of the essential functions of life. It teaches us the most appropriate food for the strong and vigorous, and directs us how to modify and re-arrange the diet of the sick and feeble. Chemistry too bears more directly than will be readily conjectured upon the life and destinies of nations. It has materials of tremendous power for the destruction of life ; yet, in its most peaceful applications, to renew and invigorate the soil, it thus tends to shed a full measure of peace and prosperity upon ages to come. It has conferred a boon of great value on mankind in the production of chloroform and chloral, substances which were once costly and of no re-cognized value, but only of scientific interest. In its products, while it has contributed much to the adorn-ment of our persons, it has also warmed, lighted, and ventilated our dwellings, purified our beverages and supplied us with beautiful utensils for our meals. While we are enumerating the boons conferred upon us by this science, the dim oriental outlines of the fabled genii rise to the recollection, by whose supernatural agencies, held in control by the magic lamp or ring, houses were built and stocked, and many other wonderful works easily performed. Such a heaven-born power is ours in the science of Chemistry, the plaything of the child, the fascination of the student, the servant of man, obedient to his bidding who has knowledge, the true talisman of power. Surely the philosopher's stone would prove a poor possession compared with the riches placed at our command by this science.

It is the intention of this work to explain the leading chemical phenomena observed in nature, and to do so, as far as possible, without the unnecessary use and encumbrance of scientific terms of expression. In carrying out this design, the simpler plan appeared to be, to treat successively the air, the earth, and the ocean ; by which means almost all that is of import- ance to be learned in the chemistry of nature, will come simply and naturally under discussion. Such a notice of the general principles of the science as is requisite to render the subsequent pages free from difficulty is added. Phenomena properly belonging to the department of Physics will here and there be noticed and explained, as all the branches of Natural Philosophy are intimately related and mutually aid in elucidating the philosophy of our globe.

SCIENTIFIC CHRISTIAN THINKING FOR YOUNG PEOPLE

BY

HOWARD AGNEW JOHNSTON, PH.D., D.D.
PRESIDENT CHICAGO CHURCH FEDERATION

SCIENTIFIC CHRISTIAN THINKING FOR YOUNG PEOPLE

Chapter I

THINKING YOUNG PEOPLE OF TO-DAY

Every year brings a host of young people to the gateway of truth, facing the task of thinking their way through the questions of life and duty. This task no one else can perform for them. A ready-made faith, without earnest inquiry, without the struggle of the soul against doubts and through questionings, may be put on like a garment; but it can never work into the life to the point of renewing the mind, and then work out as a clear conviction that becomes the compelling rule of daily living.

Religion must be real in the sense that it must be intellectually consistent with one's appreciation of values in all life, never doing violence to one's intelligence. The honest student can only believe that which he recognises as being reasonable.

Nothing is taken for granted in our day. We hear everything questioned, God, the Bible, government, and even the sanctity of the home. The whole philosophy of the social order is being attacked. The intellectual difficulties encountered by young people in finding their way to positions of confidence on many subjects are very real.

Freedom of inquiry must be encouraged. At the same time wise guidance is essential to successful study. Intelligent young people should take a thoughtful look at a rose-bud, and realise that it is not yet in flower. An open mind is a mark of a growing soul.

Only the man who has been through the struggle can be helpful to any one who is striving to build the intellectual sanctions of the Christian religion. One who has looked into some of the pitfalls, who has made mistakes himself and found it out, who has tried various paths that proved to be blind alleys, and who has come back to the main road again, can have largest sympathy with those who try to imagine that they have arrived, but know in their inmost souls that they have not.

THE SCIENTIFIC ATTITUDE IN INQUIRY

Very often a student's religious experience does not grow with his intellectual progress. A common explanation of this fact is that he is not ready to go into the laboratory of religious realities in the same spirit in which he tests teachings in chemistry and physics. Hence he often comes to feel that questions of religion are outside the realm of proper intellectual study. Later on, when he discovers that many strong men and women have a vital religion, to which he is a stranger, because of his ignorance of religious values, he hesitates to change his attitude and make an honest study of the neglected subject.

Of course such an attitude is not scientific. The

scientific attitude demands an open mind towards manifest facts everywhere, with a courageous purpose to throw over any predispositions that the new facts do not justify.

It is an unfortunate fact that many institutions of learning are not marked by a well-balanced system of teaching which guards the student against a one-sided idea of truth. Many specialists are interested only in their one specialty, and do not help their students to realise that it is but a part of a much larger whole. The result is that the total effect of the teaching is not constructive or wholesome.

Dr. Harry Emerson Fosdick knows our college life. In a recent article in the *New York Times,* he declared that we have reason for "anxious concern lest the youth of the new generation may lose that religious faith in God and in the realities of the spiritual life on which alone an abiding civilisation can be founded." He further says: "Many students are without chart or compass as far as guiding principles of religion are concerned."

TRUE EDUCATION MUST DEVELOP WORTHY IDEALS

True education must achieve a balanced training that shall not neglect the moral and spiritual nature of young people. It must lead to worthy ideals of manhood and womanhood. The training of the classroom should send out the student furnished with guideposts to indicate the way of the development of noblest character. Too many institutions furnish no such guide-posts.

Dr. John R. Mott has said: "It matters not how well

educated a man may be, if he goes out into the world with a corrupt heart, an ungoverned will and low ideals. He is a menace to society and a source of weakness to the life of his nation. Right ideals must be implanted. The springs of conduct must be touched. This is only tantamount to saying the life and principles and spirit of Jesus Christ must be brought to bear on all men individually and upon all their relationships."

That is to say, thinking is a moral activity, and must be to some purpose. Learning and science divorced from the building of character, and ignoring the value of spiritual passion, can never uplift the race. Honest doubt may be an evidence of sincere effort to find reality; but *it always seeks reality in order to live it.* Right living can never be based on wrong thinking. The honest thinker will give loyal allegiance to the best he knows.

RESPONSIBILITY OF LEADERSHIP RESTS ON YOUNG PEOPLE

Equally vital is the necessity which rests upon the leaders among our young people to accept the full responsibility of their leadership with sobered minds. They must give place to the gripping conviction of this responsibility, and accept it with courage and eager determination. This means to be thoroughgoing themselves in establishing their clear convictions regarding the supreme values in human life.

One vital suggestion is offered as a help in developing strong purpose to "make good" in realising the life that is worth while. It is to cultivate the *posi-*

tive and constructive attitude toward all subjects in the realm of inquiry, rather than the negative and destructive attitude.

Recently a young man arranged for an interview with the author to discuss the problems of religion. He asserted his dissatisfaction with his attitude toward Christianity, which was unsympathetic; yet seemed to think he was justified in it because he could not find a satisfactory ground for an intelligent faith. He mentioned some comments about religion made by men who were agnostics on the subject.

After a brief conversation, we asked him: "Do you realise that everything you have said has consisted of negations? You have not said one word to indicate a positive attitude toward anything. You must know that *negations can get you nowhere.* If you remain where you are now, you can never arrive. You know that negations lead only to destructive results in the end, and make any constructive convictions impossible. Moreover, you must realise that *a constructive program is the only one that can lead to a life worth while.*

TURNING FROM THE NEGATIVE TO THE POSITIVE ATTITUDE

He was manifestly astonished, and frankly confessed the fact, which he had not realised. He was cherishing a certain intellectual pride in his agnostic position toward things generally. We suggested that it is easy to be critical, but that *the world needs builders who are always constructors* and make all the progress that is made. He was living in an atmosphere of discontent, and could not possibly feel established in anything because of his cynical spirit.

He admitted that he had never read the Gospels with an open mind toward their message. He had never given Christ an opportunity to make His personal appeal to him. He agreed that, by every possible test, Jesus Christ must be acknowledged to be the world's greatest specialist in religious teaching, and that an honest seeker after truth must give Christ a hearing.

We urged the fact that all life takes on meaning in view of personal relationships, whether between man and man, or between man and God. We further urged that the experience of the finest men and women through the Christian centuries justified the statement that one's personal relation to Jesus Christ would determine, as nothing else could, right relations to God and one's fellow-men.

He promised to read the Gospels with these thoughts in mind. We suggested a working motto, as he was feeling his way to a positive attitude toward life. It was—*Let in the light. Let in all the light. Let the light all the way in.* Some weeks later he wrote that he was making progress in cultivating the positive attitude. Still later he wrote, with an eager note, that he had established a personal relation with Christ as his Lord and Savior. All life took on a different aspect, and is now marked by a constructive influence, as a result of Christ's life and teaching.

This positive attitude is at once scientific and meaningful. Christ gives men the truth that makes us free, an adequate philosophy of life for men as individuals and communities. To establish a right relation to him, as the world's greatest specialist in character-building, guarantees the experience of finding his truth to be the adequate light of life. In the chapters that follow we shall hope to make this statement reasonable to the candid reader.

Chapter II

WHAT IS SCIENTIFIC THINKING?

We owe a very great debt to modern science. It has opened an ever-widening horizon to the human mind. Yet many have been inclined to exalt science unduly. It should arrest our thought when we realise that what we supposed forty years ago to be accepted results of scientific research are now proved to be incorrect and untenable inferences.

We are gaining more light because we are having more science. Light is never a disturber of realities. It is always a revealer of them. No intelligent person would have less scientific research. We must have more, and still more. But we must demand that men shall be ready to accept the new light, no matter what preconceptions may be overthrown thereby.

Science has no place for prejudice, whether intellectual or religious. Science insists that mere theories shall always be held as only theories until discovered facts make them valid. Too many theories have been accepted as if they had been proved, whereas they have not been. On the other hand, many men have not been open-minded to new light, because it threatened to contravene their preconceptions. True science has suffered in both of these directions.

We may have the fullest confidence that whatever new light may come from any source, it will not be contrary to the truth which has proved its reality in

363

our actual experience. That involves the further confidence that *nothing essential to true Christianity will fail of permanency.*

THE PROPER SCOPE OF SCIENTIFIC THINKING

Unfortunately a few prominent specialists in the realm of the natural sciences, which involves nature below man, have been guilty of neglecting the study of the values to be found in the moral and spiritual realms of human life. Not only so, but they have assumed to appraise those higher values in view of their findings in their lower realms of research. Of course such assumptions have been utterly unscientific, for these men have assumed authority on subjects about which they are confessedly ignorant.

We had a striking instance of this unscientific assumption when Professor Tyndall presumed to write an essay on prayer. He admitted that he had no vital experience in the practice of prayer, and had not tested the teachings of those who were specialists in spiritual realities. Out of his ignorance he presumed to discuss a subject about which he was incompetent to speak.

We have another striking illustration of this truth. Sir Isaac Newton has been called the brainiest man in a thousand years. One day he was in a company where Professor Halley, the astronomer, was talking in a derogatory way about the Bible and Christianity. After a time, Sir Isaac said, in substance: "Halley, I like to hear you talk about astronomy and mathematics, for you have been a student in these realms. But you have made it very plain to me in this conversation that you have no right to presume to discuss the Bible and

Christianity, for you have no adequate experience on which to base an intelligent opinion. I have had that experience, which I have cultivated through many years, and because of it I am a Christian."

Here was the greatest scientist of his generation justly rebuking another scientist for his unscientific presumption in discussing a subject foreign to his actual knowledge, which comes from experience. Too many men in college faculties or in other walks of life, are incompetent to hold an intelligent opinion about the great realities in Christian experience, because they have ignored that realm of spiritual values. To ignore anything means to be ignorant of it.

Dr. Frederick F. Shannon, in a recent article in *The Christian Century*, asks this question: "Does not a large section of the educational world lay itself open to a just censure for teaching a one-sided and inadequate conception of human life? They look at one side of a proposition so constantly that they acquire the habit of mental and moral near-sightedness. Mr. Darwin himself is an example. His familiar and melancholy confession of the decay of his love for music and poetry is most saddening.

"Few generations have witnessed a deeper spiritual tragedy than that enacted by Darwin, Huxley, Tyndall and Spencer. By their monumental work on behalf of science they have made mankind their debtor forevermore. Yet they themselves were so blinded by the dust flying from the stones cut out of their scientific quarry, that they failed to give their own souls that genuine and definite spiritual opportunity for development to which they were entitled. The tragedy was all the more poignant because it was unnecessary. *These so-called educated men were terribly mis-educated men."* (Italics Shannon's.)

Thus it becomes evident that the proper scope of scientific thinking must include all the values which have been discovered in the whole range of human living. We need not argue that the highest of these values are in the realm of character-building. If we were compelled to choose between being fully informed about the facts in the realm of geology, and the actual knowledge that leads to righteous living, we would decide that the latter is of greater value to human society.

THE SCIENTIFIC METHOD IN THINKING

The thinking world has agreed that the scientific method is inductive, pragmatic, empirical, resulting from actual experiment. It is equally applicable to every subject including religion. Let us note the way it works. Every fact in the range of human knowledge is touched by a mystery. However plain the fact, the human mind cannot fully comprehend it. On the other hand, there is no mystery which to-day baffles the inquiries of the human mind, but what you will find touching it somewhere a plain, undeniable fact.

There are two possible ways of approaching the study of anything. One is to approach it at the point of its fact, to accept the fact for all that it is worth, and to continue to make the most of the fact, notwithstanding the mystery that touches it. By doing this, continually developing the values to be found in the fact, one will push the mystery further back. This is the scientific method. All the progress made in human knowledge has been made by this method. Take as an illustration the fact of electricity, touched by its mystery. It has been by making the most of the fact that we have pushed its mystery further back, thus increasing our knowledge of the fact.

But there is another way of approaching the study of anything. That is to approach it at the point of its mystery, and refuse to acknowledge the fact because we do not understand its mystery. Of course this is not scientific, and it is not honest. Any one actually desiring light in any realm will not refuse to accept the manifest facts to be had, with their values.

Yet this is what many people do in matters of religion. In fact, it is only in religion that any one ever adopts this method. No one ever refused to accept the fact of electricity, with its light and heat and motive power, because he did not understand its mysteries, which still baffle Edison and Marconi. This attitude is never justified by intellectual honesty.

CERTAINTY DOES NOT DEPEND ON COMPREHENSION

It follows that we do not need to comprehend anything fully in order to be certain of it. We may apprehend it, though we do not comprehend it. It follows also that it is not necessary to explain a fact of experience in order to be certain of it, and to declare our sense of certainty with confidence. Electricity will serve again as an example. Such confidence is entirely reasonable in every realm.

This is what we find in connection with the experiences of religion. Many are certain of the facts in religious experience, though they cannot fully explain them, especially to some one else who has never entered the realm of that experience. It would be equally difficult to explain the facts about electricity to one who had never given attention to the subject.

Having this scientific method in mind, let us emphasise the truth that the greatest fact in the world is life; also that the greatest mystery in the world is life. Moreover the greatest life in the history of the world is Christ. What, then, will the scientific method require of us in our attitude toward Jesus Christ? The answer is immediate. If we be scientific, we must accept the fact of the life of Christ, notwithstanding any mystery that may attach to that life.

We must make the most of the facts involving the amazing influence which the character and teachings of Christ have exerted, and now exert, upon human history. We must continue to develop all the possibilities which lie within the range of the life of Christ, and thus push the mystery of Christ further back. Thus we will come into right relations with him, and will possess more and more the values which that relationship secures. This is scientific thinking as it applies to the realm of supreme values in human living.

CHAPTER III

CHRISTIAN THINKING NECESSARY

In recognising the necessary place of religion in human life, the scientific inquiry must be as to which of the religions known to men is worthy of first place in our appreciation. That will be the religion which has given the highest spiritual values to mankind.

Without indulging in any invidious comparisons, gladly acknowledging certain values in other religions, frankly confessing the failures of many Christians to exemplify faithfully the truth of Christianity, without evading any fact whatever that may have any bearing on the question, we confidently assert that no candid student of history can question that Christianity must be given first place.

The explanation of the supremacy of Christianity among the religions of men is the character and teaching of Jesus Christ. He has won the high place of being recognised as the greatest specialist in religion known to men, both in the quality of his teaching and the character of his life. Hence the religion which we exalt in this book is Christianity.

THE MATCHLESS CHARACTER OF CHRIST

Mr. Lecky, in his *History of European Morals,* declared: "It was reserved for Christianity to present to the world an ideal character which through all the changes of eighteen centuries has inspired the hearts of men with an impassioned love; has shown itself capable of acting on all ages, nations, temperaments and conditions; has been not only the highest pattern of virtue, but the strongest incentive to its practice. The simple record of those three years of active life has done more to regenerate and soften mankind than all the disquisitions of the philosophers and all the exhortations of the moralists. This has been the well-spring of whatever has been the best and purest in the Christian life." Quotations of like import might be multiplied indefinitely.

Chief Justice Taft is known to be a man of broad and liberal culture, and candid to a degree. Soon after his return from the Orient, he delivered an address in Carnegie Hall, New York, on the importance of assisting the enterprise of Christian missions to non-Christian peoples. In that address he said: "No man can study the movement of modern civilisation from an impartial standpoint and not realise that Christianity and the spread of Christianity are the only bases for hope of modern civilisation in the growth of popular self-government. I think I have had some opportunity to know how dependent we are on the spread of Christianity for any hope we may have of uplifting the peoples whom Providence has thrust upon us for guidance. I did not know until I went into the Orient. In the progress of civilisation you cannot overestimate the importance of Christian missions."

The moment we assert that Christ is the dynamic in Christianity which explains its superiority over all other religions, the question will be asked: "Which Christ?" A dozen Christs emerge at the mention of his name. There can be but one answer—the Christ of the Gospels. Let all ecclesiastical non-essentials be relegated into the background, and let the Christ portrayed in those Gospels make his own appeal. He will vindicate his claim to be "the way, the truth and the life."

When Sherwood Eddy gave his testimony as an ambassador of Jesus Christ to several groups of officials in China, he asked them, in all fairness, to study the Gospels for themselves and get acquainted with Christ at first hand. Many responded to that reasonable request, and on his next visit to China Eddy found many of them confessing Christ as their Savior and emphasising their conviction that Christianity is the only adequate religion for mankind.

CHRIST'S TEACHING MEETS EVERY HUMAN NEED

Prof. Francis G. Peabody, of Harvard, in his book *Jesus Christ and the Christian Character*, makes this inviting statement to all inquirers: "When one turns to the Gospels, he discovers with fresh surprise the extraordinary richness and variety of the teaching of Jesus. *Each period in history goes with its question to the simple record, and finds an answer* which seems written to meet the special problem of the time." Let those who tell us that modern demands for leadership have outgrown Jesus Christ ponder this declaration of the Harvard professor, who finds Christ adequate for every such demand.

Coming closer to his reader, Dr. Peabody says: "The teaching of Jesus, even when its form is social, is fundamentally personal. Out from behind the social question emerges the antecedent problem of the Christian character. What are the traits which Jesus is most concerned to inculcate? By what kind of persons is the service of the world to be effectively undertaken? Is the character trained in the way of Jesus fit to meet the demands of the present age?"

Note the answer which our author gives: "Such an inquiry would seem to be peculiarly free from difficulty. It appears to lie on the very surface of history, and to require no venture into the depths of criticism or speculation. Nothing would seem to be more easily determined than the kind of character which is inspired and exemplified by Jesus Christ. A return to the teaching of Jesus is essential if Christian ethics is to have a hearing from this present age."

THE MAJORITY OF LEADING SCIENTISTS HAVE BEEN CHRISTIANS

Since the Renaissance, when modern science had its beginnings, the vast majority of leading scientists have been Christians. The popular mind has not realised this fact, because five or six prominent scientists have not been Christians. From the days of Copernicus and Galileo, of Kepler and Bacon, of Newton and Kant, down to our time, most of the men who have made real contributions to scientific knowledge have been loyal followers of Jesus Christ. Among them may be mentioned Hugh Miller, Agassiz, Linnæus, Livingstone, Virchow, Lord Kelvin, Dana, Gray, Pasteur, Liebig, Romanes, Ampere, Faraday, Mendel, Maxwell, LeConte and many more who might be mentioned.

The logic of the foregoing statements and testimonies is clear. There is a new day awaiting the sincere seekers of the solution of the problems to be found in the pathway of human progress. It is the day of Jesus Christ. He is the supreme teacher of the truth that vitalises human life with that spiritual quickening which is the soul of the highest culture. His way is our hope. The actual experience of history points to him as to no other, as we seek an adequate program involving all the relationships of men.

THE WORLD NEEDS CHRIST'S LEADERSHIP

It would be difficult to find a more cogent statement of the need of the leadership of Christ in the life of our time than was made in a recent article in *The Continent* by Dr. William Pierson Merrill, of New York City. He said: "The world waits in its desperate need, crying out for a Savior. And it will wait until we Christians are ready to take as our sole and sufficient faith the conviction that Christ and Christ only is the Savior of the world; until we get into the temper that does not care much about anything except to get him, his way, his ideals and his spirit into the life of the world in all its phases."

He declares that believers must be faithful and fearless in saying "to business leaders and working men and social theorists and all the world, that the only way out is the way of Christ, the way of frankly accepting Christ's ideals as the way to do business, putting service in the place of profits as the standard of success, putting and keeping personal and human interests above all property interests, humanising and Christianising the whole business and industrial order. . . . You are doomed unless you take the way of Christ, at any cost."

Of the larger field of Christian conquest, Dr. Merrill said: "No league or association of nations will ever work, or be more than a dream on paper, unless there come through all nations a great spirit of good will, of mutual respect, of patient consideration in their dealings with each other. And what is that except to say that there is no salvation, no durable peace, except through Christ? His spirit must rule, that spirit which is the antithesis and the denial of racial hatred, of national selfishness, of suspicion and distrust, which thwart the fairest plans. There is no salvation for the international order except in Jesus Christ. All through the world's life there is an impotent waiting, an inarticulate longing, for the force that can save the world out of its distress. And the only answer is Christ."

CHAPTER IV

THE EXERCISE OF FAITH UNIVERSAL

By faith we do not mean credulity. Some people seem to think that religious faith is a sort of blind acceptance of teaching, without the ability to justify that acceptance by vigorous intellectual sanctions. There is an intelligent Christian faith which is actually scientific in its character and exercise, and which is the inevitable possession of those who faithfully and fearlessly face all the light that may be brought to bear upon the big questions of religion which demand investigation and decision.

This intelligent faith is so grounded in actual knowledge, in personal experience, in repeated testings, that it often goes beyond the point of clear and strong convictions, and carries in it a sense of certainty that cannot be shaken by doubts or fears.

Yet when we come to discriminate between faith and demonstrated knowledge, it must be kept in mind that faith supplements knowledge and rests upon it. Thus the apostle John, in explanation of his faith in Christ, asserts in the beginning of the first letter which bears his name: "That which we have heard, which we have seen with our eyes, which we have looked upon, and our hands have handled of the Word of life; that declare we unto you, that ye also may have fellowship with us." The certainty of actual experience is the basis of his intelligent and necessary faith.

THE EXERCISE OF FAITH UNIVERSAL

HUMAN KNOWLEDGE IS LIMITED, HENCE FAITH

There is a superficial notion that faith is peculiar to the realm of religion, whereas it is being exercised in every walk of human life. The necessity for the constant exercise of faith is found in the fact that the human mind is limited in its capacity. It is hung up between the infinite and the infinitesimal. No man can imagine the end of space, for he could immediately think of going beyond that point.

Nor can we imagine the infinitesimal, for you can think of the smallest particle of matter being half as large. It is because of this limited character of the human mind that we must supplement actual knowledge constantly with inferences, beliefs, which reach beyond the point of knowledge, but which rest upon what we do know. Moreover, this is intelligent faith, a reasonable exercise on the part of men.

WE EXERCISE FAITH EVERYWHERE

We are told the business world transacts ninety-five per cent of its business on paper, which is a series of "credentials," indicating faith on the part of people who do business with each other.

A happy home could not exist without faith. Too often it is impossible to maintain because of suspicions and betrayals, until faith is gone, and the home life is shattered. In every case intelligent faith is based on actual knowledge and experience.

In the realm of natural sciences there is a vast exercise of faith, to which we must make special reference. We find no fault with it, for it is as unescapable there as everywhere else. But we must realise that it is the same kind of faith that is exercised in matters of religion.

Take the science of geology, where specialists have given out various statements about their belief concerning the age of the earth. No two agree. Positive proof is impossible. Each man based his conclusions, his beliefs, on certain data, which he felt would justify his inferences. It was a matter of faith, a justifiable faith, yet faith.

Or take the science of physics. Some physicists tell us there is no such thing as luminiferous ether. Most scientists think there is. There is no final demonstration possible. It is a matter of faith. Remember, we do not object to it. We cannot get along without it.

Or take biology. Darwin believed for years that the so-called process of natural selection, in the transmission of persistent variations through ages of time, explained the origin of species in plants and animals. Hundreds of students of science have believed it, mainly because Darwin said so. It is a fine instance of faith, without personal investigation. But Mendel's law, as we shall see in a later chapter, has proved that natural selection is utterly untenable, and leading scientists have discarded it. Here was a faith bordering very closely on credulity.

Now suppose a man who is one of the greatest leaders of the world's life should take his place beside these scientists. For instance, a man like Gladstone, of whom Lord Salisbury said that he was not only a great financier, not only a great scholar, not only a great statesman; but also and most of all a great Christian. Suppose Gladstone should say to them: "Gentlemen, I too believe something, which is based on the most real experience of my inner consciousness. I believe the life and teaching of Jesus Christ have wrought into human life the most powerful transforming influence ever known to mankind. I am compelled to believe that the change which Christ has wrought in my own life justifies my position. It is my unshaken belief that Christ brings to mankind the greatest blessings ever offered to human beings in the realm of spiritual values."

What must we say of the reasonableness of Gladstone's statement? There is only one possible answer. If we be true to scientific principles, on which intelligent faith is based, we must say that Gladstone's faith is a reasonable, a logical, a scientific faith, an unescapable conviction resting upon the most vital experience of reality he ever knew. And exactly this faith is possessed by thousands of intelligent and sincere Christians.

FAITH IS POSITIVE AND CONSTRUCTIVE

It is very important to recognise the fact that faith is always positive and constructive, and therefore certain to be helpful. Doubt is negative, and cannot contribute to stability. Many people will tell you they cannot believe various things for various reasons. In later chapters we shall consider many of the problems which unbelievers present for solution. The point to be noted here is that these people, in their negative position, have no philosophy of life which they can offer as a better substitute for the Christian position. They have no solution to offer to life's problems that will bring light and hope, comfort and peace, courage and strength to the soul.

Yet these same people often think of themselves as being broad-minded, and of Christians as being narrow. But what is the fact? No one can be broad-minded who does not have a philosophy of life, which takes all the facts into account, and offers light upon our way. The Christian philosophy of life is big enough to cover all the facts in human experience. Hence the Christian is broad-minded as others cannot possibly be. It is the unbeliever who is narrow and superficial, when he comes to solutions for the problems of the immortal soul. His negations are helpless to bring satisfaction.

The reason for the Christian's breadth of vision, his wide-spread sympathy with every fellow-man who needs light on his pathway, is his *experience of the fact* that Jesus Christ has shown mankind the way out of the darkness into light, the ability to meet him at every point of need with help and strength, with cleansing and victory. He realises that *Jesus Christ is the broadest-minded man that ever lived.* Hence he gives open allegiance to this sufficient Savior and his program for the redemption of the world unto the fellowship of God.

CHAPTER V

THE CONVERSION OF PROFESSOR ROMANES

Several of the statements made in the preceding chapters find illuminating illustration in the striking experience of Prof. George J. Romanes, of the University of Cambridge. He held the chair of Biology in that university and was editor of the periodical entitled *Nature*. His influence was great in the realm of the natural sciences, and he was active in current discussions of the evolution problems. He was called the greatest Darwinian after Darwin.

Early in his career Professor Romanes published anonymously a little volume entitled *A Candid Examination of Theism, by Physicus*. In that book he said he could not believe in the existence of God, nor in the immortality of the soul, because he could not demonstrate the reasonableness of either view to the satisfaction of his intellect. He felt compelled to take his place with the agnostic materialists and atheists.

But he was not satisfied. Very frankly he confessed that he had no such happiness in his unbelief as he once had in his faith in God and a spiritual interpretation of the universe. He was thoroughly candid, and was seeking for light which would restore to him the satisfaction he had lost.

One day, while using his microscope, he suddenly stopped. The idea occurred to him that physical forces have an intelligent power back of them which explains their movements. He asked himself this question: "Since science demands experience as the basis for intelligent conviction, is there anywhere in the range of experience the presence of an intelligent power that is an adequate explanation of the movement of physical forces everywhere apparent?"

THE SPIRIT NATURE OF THE HUMAN WILL

Immediately Romanes thought of the human will, and of the self-conscious, intelligent, volitional being, who finds expression of that will in the control of physical forces. I decide to lift up this book, and the book comes up. Why? Just because I willed to do so. Then Romanes recalled that Alfred Russell Wallace had emphasised the fact that the will is of the nature of spirit, and that it is of the nature of spirit to control physical forces.

The great realities in any company of people are not anything you can see or feel. They are the invisible spirits in communion with each other, as they give expression to their presence and thought and power. You should never say, I have a soul. You should always say, *I am a soul.* Moreover, the spirit reaches far beyond the basis of his activity in the body, along the lines of the media of communication and activity with which he is familiar. That familiarity now involves far more than was known a few years ago. We sit down in Chicago and talk to Boston or San Francisco, and a thousand miles are annihilated in an instant.

With some such thoughts Romanes visualised the significance of the fact and nature of the human spirit. He quickly realised that he had not been scientific in calling himself a student of nature, for *he had been ignoring the most important part of nature, namely, human nature.* He had been studying nature below man, and had been judging the higher realm by the knowledge of life he had acquired on the lower levels. It is never justifiable to interpret man by the creation below him; yet this is exactly what many natural scientists have done.

Romanes also saw that the uniformity of law and the solidarity of the universe compelled him to infer that back of all physical forces there is an intelligent, volitional being of the nature of spirit, who controls the universe of worlds and determines their processes. This Being must possess unspeakably vaster reach of thought and power of control.

Then this great biologist began to study man, as he had not done before. Very soon he was profoundly impressed by the evidence of the religious instinct in man. Biology had taught him that whenever instinct appeared in any form of life, it always pointed to something which satisfies the instinct. Romanes again accepted the necessity involved in the uniformity of law, and became convinced that the religious instinct in man pointed to moral and religious attributes in the being who ruled the universe.

Thus he came to what is called the theistic position, *compelled to the belief in God by his fidelity to the scientific method.* That was in the year 1890. In that year the magazine *Nature* carried a discussion about various phases of the theories of evolution. One of the contributors to that discussion was Dr. John T. Gulick, of Osaka, Japan, a Christian missionary with decided ability as a naturalist. Romanes said of him:

"He brings the most profound intellect to the discussion of the subject."

ROMANES INVESTIGATES CHRIST

On Christmas day, 1890, Romanes wrote to Gulick: "I have long wanted to ask you a question, for two reasons: First, because I know a man of your intellect would not believe anything without a good reason; second, because I know you would not profess to believe anything without being sincere. I wish to know *how you can believe in Jesus Christ as the Savior of the world.*"

Dr. Gulick's answer to that question is published in the *Bibliotheca Sacra,* for April, 1896. Two points he stressed. He said: "I ask you to approach the subject with me from the viewpoint of biology. The science of biology always recognises a new type of life because it exerts an influence upon its environment different from any other known. *Jesus Christ has exerted an influence upon his environment different from that which any other type of life has exerted.* Biology must take all life into account, and must recognise Christ as a new type of life in the world. Study the history of humanity, as it has been influenced by Christ, and we cannot escape the conviction that he reveals a type of life unique and unparalleled."

Romanes was impressed even more powerfully by Gulick's second point. He said, in substance: "I am afraid that you have made the mistake that many men make of supposing that the intellect is the only organ of evidence to the soul. The affections are also an organ of evidence to the soul. and the will is an organ of evidence to the soul, especially in the realm of personal relations. While this evidence is never contrary to the intellectual appreciations, it is distinctive in revealing truth and reality which the intellect cannot reveal."

Romanes declared that the day this truth broke upon his appreciation was the greatest day of his life. He said: "I believe in the Copernican theory in astronomy because I can demonstrate its reasonableness by mathematical processes. But *I know that my mother loves me, and I cannot prove it by logic.*" He also said: "I had never taken seriously that saying of Jesus, 'If any man willeth to do his will, he shall know of the teaching whether it be of God'; but I find that Jesus was scientific, in that he gave us a working hypothesis in that saying, challenging us to test and prove that it works. No man ever tried it who did not prove it true."

Then Romanes began to study the Gospels, with a fascination which he had never thought possible, and with an eager open mind. As he measured the whole sweep of the amazing influence which Christ has exerted upon humanity, he steadily came to believe that the truth which Christ has given the world is the only adequate light of life and hope of men. Then he revealed his unhesitating courage of conviction and sincerity of character by uniting with the Church of England as a humble Christian.

THE TRAGEDY OF WRONG LEADERSHIP OF YOUNG PEOPLE

Of one thing Romanes was keenly regretful. During the days that he had not been truly scientific, ignoring the vital values in the Christian religion, he had exerted a great influence upon his students. They said: "There is the great Romanes. He is not a Christian. Why should we be concerned about these things? Romanes is the spokesman for science, and he leaves religion out of account."

Being determined to do what he could to counteract this hurtful influence, Romanes was gathering material to write another book, when his untimely death interrupted his plans. His friend, Canon Gore, then of Westminster, put the material into a volume and published it under the title *Notes on Religion, by George J. Romanes.* That book contains many of the facts reported here.

The experience of this great scientist reinforces the challenge that Jesus Christ makes to every intelligent man and woman to study his life and teachings with an open mind, and test his claims in the laboratory of daily life. Many others know the dissatisfaction which Romanes confessed in his agnosticism. He very earnestly declared that if any man will be absolutely honest with all the light that shines upon the subject, and will be faithful to the scientific spirit and method, he will inevitably be led to the feet of Jesus Christ as the Savior of the world.

Chapter VI

THE VALUES OF EXPERIENCE

Modern science rightly stresses the necessity of the laboratory method. The experience resulting from actual experiments must be the basis of all intelligent convictions and beliefs. Only thus can we justly determine what values are worthy of adoption in our program of daily life.

The greatest of laboratories is the realm of man's daily living. There we must test the claims of all who seek our allegiance to their leadership along various paths of activity, which they would have us believe to be paths of progress. The one test which must determine our final decision regarding any subject is *the way it works in daily life.*

It was in recognition of this basis of judgment, as inevitable, that Jesus said to his disciples: "By their fruits ye shall know them. Men do not gather figs of thistles." Time reports the results of our testing. Ideals proposed by various leaders are thus tested, and cast aside as inadequate, or adopted as worthy of acceptance.

THE RELATION OF EXPERIENCE TO AUTHORITY

Recognised authority always involves obligations on the part of those who accept it. Sometimes people are subjected to power which assumes authority, without their consent. But in time, when it became evident that unjust conditions followed, the people have repudiated it.

When Jehovah gave the commandments to Moses for the people of Israel, he did not arbitrarily assert his authority to do so; but reminded them that he had brought them out of Egypt and its bondage. He rested his claim to their allegiance on the fact that he had proved to be seeking their betterment, and on the favourable presumption, which their experience justified, that his way is the best way leading to increasing blessings.

Dr. Ernest F. Tittle, in the *Christian Century* of Sept. 21st, 1922, wrote: "Very great indeed is the authority of Jesus to-day. Slowly but surely the world is opening its eyes to the fact that what is ethically un-Christian is economically and politically unsafe. But if men believe, as many of them are beginning to do, what Jesus said, it is because what Jesus said is being verified by the accumulating experience of the race.

"And what is true of the authority of Jesus is true, likewise, of the authority of the Bible and the authority of the church. If certain teachings of the Bible, certain pronouncements of the church, are being accepted to-day, it is not merely because they are found in the Bible, or have been uttered by the church, but only because human experience is showing more and more clearly that they are true."

NOMINAL ACCEPTANCE OF AUTHORITY IS VALUELESS

Consider a suggestive fact which bears upon this discussion. Some decades ago there were two groups of men who lived on different sides of Mason and

Dixon's line. They agreed about the final authority of the Scriptures. But one group said the Bible taught that slavery is right, while the other insisted that the Bible taught slavery to be wrong. Agreement as to authority did not bring agreement as to rules of conduct. Who would decide as to the teaching of the Bible on the subject of slavery?

Time has brought a growing consensus of man's moral judgment that slavery is wrong, and that the Bible condemns the institution. It is an Old Testament institution, and the New Testament sets aside much of the Old Testament. Nothing in the teaching of Christ encourages slavery; but that teaching contains much that will make slavery impossible. Moreover, as Christ's teachings are increasingly valued by men, in the light of experience, the application of his principles of righteousness will make all slavery impossible, economical, industrial and intellectual, as well as spiritual.

Consider another fact. If a Christian in Constantinople, with his Bible under his arm, should meet a Mohammedan who would ask him what book he had, he would reply: "This is the record of God's revelation to men." To which the Mohammedan would say: "Oh, the Koran!" Quickly the Christian would reply: "No, not the Koran. This is the Christian's Bible." But the Mohammedan would declare that the Koran is the true revelation of God to men. Evidently no progress would be possible on the ground of asserting each man's authority to be the correct one. Only as the teachings of the two books reveal their values in daily life can the advocate of each make headway with his claims.

SCIENTIFIC CHRISTIAN THINKING

RELIGION THE DOMINANT FACTOR IN HUMAN LIFE

The verdict of history is the recorded finding of human experience. History has proved that the determining factor in the life of any people is not the intellectual factor, not the social factor, not the commercial factor. It is always the moral and religious factor. Every candid student of China will agree that Confucianism explains the life of China as nothing else could. The fullest expression of Buddhism is in Burmah and Siam, and has had more to do in shaping the life of the peoples of those countries than anything else.

Every one who knows India cannot doubt that Hinduism saturates the life of that country, with its system of caste and its atmosphere of pantheistic religio-philosophy, determining India's life as nothing else has. No less evident is it that Mohammedanism is the supreme explanation of the Turkish empire as it existed for more than a thousand years.

It is equally true that Christian countries have been quite as clearly developed by that type of Christianity which has dominated the lives of the people as by no other influence. Russia and Greece have been the response to the teachings and practices encouraged by the Greek Church. In like manner the teachings and practices encouraged by the Church of Rome have determined, as nothing else, the life of France, Spain, Italy, Mexico, Central and South America. True also that the Protestant countries of Great Britain, Germany, Holland, the Scandinavian peoples, the United States and Canada have developed distinctive character as a result of that Christian teaching and practice which gave the people the open Bible and emphasised direct access to God.

Experience has also proved that the final evidence of the value of any religious teaching must be found, not in the number of adherents, but in the quality of character manifest in the daily life. Furthermore, if certain countries have given religious teachings larger place than others, it follows that one has a better opportunity to test its real value in that country which gave it the most loyal response.

Thus it is fair to say that France has not given as large a response to the influence of the Church of Rome as Spain has done, so that it would be just to consider the religious life of Spain as more indicative of the real influence of Romanism. In like manner it is evident that Germany has not given as large a response to Protestantism in the last hundred years as the United States has done. Hence it is just to turn to the United States to discover the fruits of Protestantism.

Manifestly there is real difficulty involved in making a just and discriminating test of the true values of any religion, in view of these varied degrees of response given to it by the people to whom it has made its appeal. Yet, taking into account every possible fact that will bear upon the subject, including the results of the efforts of any religion to carry its influence into the realm of other teachings, we make the claim without hesitation that the superiority of Christianity has been demonstrated by the religious experience of those who know it best, as compared with any other religious teaching.

Chapter VII

SCIENCE AND EVOLUTION

In the periodical *Science,* for April 14th, 1922, Prof. Wm. E. Ritter, of the Biological Research Department of the University of California, writes concerning discussion on the subject of the present status of the evolution theory: "If one scans a bit thoughtfully the landscape for the last few decades, he can hardly fail to see signs that *the whole battle-ground of evolution will have to be fought over again;* this time not so much between scientists and theologians, as among scientists themselves."

Prof. Thomas Hunt Morgan, of Columbia University, in his recent book, *A Critique of the Theory of Evolution,* tells us: "To-day the theory of natural selection has few followers among trained investigators, but it still has a popular vogue that is wide-spread and vociferous."

Dr. Wm. Bateson, Professor of Biology in Cambridge University, in his recent book, *Mendel's Principles of Heredity,* after a full discussion of the revolutionary overturning of the generally accepted theory of natural selection, as a result of Mendel's law, as proved by long and careful personal investigations and testings, confidently asserts that if Darwin had known the truth revealed by Mendel's law, he would never have written his books.

Some years ago Dr. Wm. Carruthers, Curator of the Botanical Department of the British Museum, then the retiring President of the Linnæan Society, told the author that he was certain that Agassiz would be vindicated within a few decades for having rejected the theory of evolution, as being without adequate proof to justify it; and that *science would be compelled to find some other explanation of the processes of nature.*

For some years the author accepted the theory of evolution as probably the correct statement of the method which the Creator had adopted in creation. He joined the growing ranks of many who considered the prevailing opinion of scientists one to be followed, especially as he found no difficulty in recognising the fact that if God had adopted evolution as the method, it in no way interfered with his established convictions about God as Creator and Ruler of the universe.

The conversation with Dr. Carruthers, reported above, led him to make a more careful study of the actual facts which scientists had found in their investigation of the evidence to support the theory. Watching the reported results of the continued study of available facts by leading scientists, he has become convinced that Dr. Carruthers is justified in his judgment that scientists will discard the theory of the organic evolution of species, and turn to some other explanation of the processes of nature.

THE THEORY STATED

According to this theory, plants and animals, as individuals, develop a variation which was not found in the parent. This variation is very slight, often imperceptible, thus requiring ages of time for its definite formation. But it persists because of its usefulness, and is reproduced by transmission to offspring, until it becomes so marked that it produces a new *variety* of plant or animal. This variety continues to become more pronounced until its differentiating features justify one in calling it a new *species*.

It is well known that very often new species were announced which were only varieties, although marked by wide variation. A new species to be recognised by science *must show some new character which no ancestor possessed,* and must show that this new character *will breed true in all circumstances of hybridisation, and persist through continuous transmissions.* There must be *a difference of form, structure and habit to* constitute a new species.

If all the young of the different species were to grow to maturity, there would not be food enough to keep them alive. Hence Mr. Darwin, in his book, *The Origin of Species,* emphasised the constant struggle for life on the part of all living things. Popular usage calls this *the struggle for existence.* Darwin held that the offspring favoured by this new variation, however slight, could better secure food, or withstand an enemy, so that it would survive where others perished. Herbert Spencer called this *the survival of the fittest.*

Darwin held that *"probably"* nature is constantly selecting the forms best calculated, through adaptation, to compete with other organisms for existence, and originated the expression *natural selection* to describe this process. He claimed that "natural selection acts solely through the preservation of variations, in some way advantageous, which consequently endure."

The offspring reveals one of two tendencies, either to resemble the parent or to differ from the parent.

Darwin held that some offspring develop the tendency to differ from the parent so strongly as to carry the variation off into new forms, which never return to the ancestral form. It is this particular point in Darwin's theory that has been proved to be incorrect.

It is very important to note that evolutionists assume ages of time as necessary to realise the evolution which they advocate. Darwin says: *"We see nothing of these slow changes in process,* until the hand of time has marked the lapse of ages; and then so imperfect is our view into long past geological ages, that *we see only that the forms of life are different* from what they formerly were."

Here Darwin admits that *no actual processes of evolution are in evidence,* and no new species are being produced before the eyes of investigators. Hence he must claim that such evolution took place ages ago, long before even the fossil remains existed. For *the fossil remains show no gradual changes from one species to another.*

A further fact should be mentioned. Evolution has involved the idea that the creative agencies at work in the world are indirect, rather than direct. But the findings of scientists, especially the latest discoveries and demonstrations, point to the evidence of a much more direct contact on the part of the Creator with his creatures.

TWO GROUPS OF EVOLUTIONISTS

Among the evolutionists there has been a group of materialists. In a former chapter we noted that these men have ignored certain realms of vital realities and supreme values. Their views have been destructive of Christian teachings at various points, as we shall note later. In recent years their position has been widely discounted.

A second group is made up of scientists who have also been Christians. They had already established their convictions concerning God and his vital relation to men and the whole creation. Hence any new light on the method which God employed in the processes of nature could not change their convictions, which were based on actual spiritual experience. They knew that truth anywhere would harmonise with truth everywhere.

No fair-minded student of the whole subject can be justified in saying that many evolutionists were not also sincere Christians.

EARLY OBJECTIONS TO EVOLUTION BY SCIENTISTS

Before Darwin was prominent, a French scientist, Jean Baptiste Lamarck, advanced the general theory of evolution. At various earlier times suggestions had been made of something of the kind; but Lamarck was the first to develop the theory distinctively. Scientists began at once to examine the available evidence to determine how far there were any facts to justify the theory.

One of the most careful and competent of these was Hugh Miller, whose brilliant work, *The Old Red Sandstone*, remains a classic to this day. Concerning Lamarck and his theory Miller wrote: "The ingenious foreigner, on the strength of a few striking facts, has concluded that there is a natural progress from the inferior order of being toward the superior. *He confounds gradation with progress. Geology furnishes no genealogical link to show that existences of one race derive their lineage from the existences of another."*

Miller cites a striking proof in the case of fishes. He says: "Of all the vertebrata fishes rank lowest, and appear first in geological history. Now fishes differ among themselves. Some rank nearly as low as worms, some nearly as high as reptiles. If fish could have risen into reptiles, and reptiles into mammalia, *we would necessarily expect to find* lower orders of fish passing into higher, and taking precedence of the higher in their appearance in point of time. But it is a geological fact that *it is fish of the highest order that appear first on the stage,* and that *they are found to occupy exactly the same level during the vast period represented by five succeeding geological formations. There is no progression, and the argument fails."*

Prof. Joseph LeConte, of the University of California, laid great stress upon the fact of missing links being "the greatest of all obstacles to the theory." In his book, *Religion and Science,* he declares: "The evidence of Geology to-day is that species seem to come into existence *suddenly and in full perfection,* remain substantially unchanged during the term of their existence, and pass away in full perfection. Other species take their places, apparently by *substitution, not by transmutation."*

Darwin himself confessed this lack of evidence to support his theory. In his *Life and Letters,* he said: "I do not pretend that I should ever have suspected how poor was the record in the best preserved geological sections, had not the absence of numerous transitional links between the species which lived at the commencement and at the close of each formation, pressed so hardly upon my theory."

Prof. Huxley, in his *Lay Sermons,* admits that *an impartial survey of the positively ascertained truth* "either shows no modification, or demonstrates it to have been very slight, and yields *no evidence whatever that the earlier members of any long-continued group were more generalised in structure than the later ones."*

Prof. J. Arthur Thompson, of Aberdeen University, admitted to be one of the highest living authorities, in his book, *Heredity,* Revised Edition, 1919, says: *"The question resolves itself into a matter of fact.* Have we any concrete evidence to warrant us in believing that definite modifications are ever, as such, or in any representative degree, transmitted? It appears to us that we have not."

With leading scientists thus casting positive discredit upon the evolution theory, one is at a loss to understand how it ever gained such headway. Darwin himself confesses his amazement at this phenomenon. In his *Life and Letters,* he says: "I was a young man with unformed ideas. I threw out theories and suggestions, wondering all the time over everything, and to my astonishment the ideas took like wildfire; people made a religion of them." This candid statement suggests that men liked the theory, whether it could be proved or not. But theories without validating proofs cannot stand.

SPECIAL THEORIES IN EMBRYOLOGY

The first conspicuous critic of Darwin was Prof. August Weissmann, of Germany, a gifted biologist.

He was an admirer of Darwin, but his earnest study convinced him that Darwin's theory was wrong. Weissmann contended that the germ plasm is the basis of heredity. To put it in a homely phrase, he held that the eggs (the generative cells) determine the form of the hens, rather than the hens the eggs. *He denied Darwin's view that individual offspring developed variations that persisted and were transmitted, declaring that no proof to justify the theory could be found.*

Prof. Ernest Haeckel followed Weissmann's ideas. His name is identified with the *biogenetic law* in embryology, which is also known as the *recapitulation* theory. According to this theory, the developing embryo of a higher animal passes through the various successive stages which marked the embryos of lower forms of animals, thus seeming to indicate that the process of evolution had involved these lower forms in its progress through them to the higher form. This theory made the argument for evolution very impressive for a time.

Haeckel produced a set of diagrams of supposed embryo fossils, showing the "missing links," and claimed to have solved the "riddle of the universe." But careful scientists could find no evidence of fossils which justified the diagrams and claims of Haeckel. He was accused of forgery, and tried by the Jena University court. By his own confession, which appeared in the *Muenchener Allgemeine Zeitung,* some of his drawings were purely fictitious.

Haeckel said: "I begin at once with the contrite confession that a small per cent of my embryo diagrams *are really forgeries,* those namely for which the observed material is so incomplete or insufficient as to *compel us to fill in and reconstruct the missing links by hypothesis* and comparative synthesis. . . . I should feel utterly condemned by the admission, were it not that hundreds of the best observers and most reputable biologists lie under the same charge. The great majority of all morphological, anatomical, histological and embryological diagrams *are not true to nature,* but are more or less schematised, doctored and reconstructed." Yet Haeckel's diagrams are taught in many schools to-day, as if they presented facts.

Extensive tests of the theories of Weissmann and Haeckel have proved that *the embryonic development of certain animals has nothing in it whatever to suggest a recapitulation* of that animal's history.

Probably the latest word on this subject is from the pen of Prof. A. Weber, of the University of Geneva, who writes in the *Scientific American Monthly,* for Feb., 1922. In an article entitled *The Mechanical Side of Evolution,* he dwells at length on "some remarkable recent discoveries in the field of embryology," and says: "It was long ago clearly demonstrated by palæontologists that many embryologists had confounded the phylogenetic development of species with that of the organs considered separately. The success of the experiments undertaken in order to verify Mendel's laws of heredity have been of no little help in creating an entirely new attitude among embryologists."

Prof. Weber then adds: "The critical comments of such embryologists as O. Hertwig, Keibel and Vialleton, have practically torn to shreds the aforesaid fundamental biogenetic law. Its *almost unanimous abandonment* has left considerably at a loss those investigators who sought in the structure of organisms the key to their remote origins or to their relationships." Thus what was the most impressive argument for evolution for many years has been set aside by leading embryologists.

DARWIN'S IDEA OF THE STRUGGLE FOR LIFE INADEQUATE

For a time Darwin's idea of the struggle for life was generally accepted. Its effect was depressing. It tended to develop an atmosphere of pessimism by teaching that the struggle for existence is hard and cold, its outcome determined by sheer physical strength. Higher values were ignored. Henry Drummond attacked this theory most vigorously in his *Ascent of Man*, and made a real contribution to the general subject. Drummond declared that this theory "misread nature itself, in fixing upon a part whereby to reconstruct the ultimate, which is not the most vital part, and therefore the reconstructions have been wholly out of focus."

While fully recognising the struggle for life, Drummond says: "But that it is the sole, or even the main agent in the process must be denied. There is a second factor which one might venture to call *the struggle for the life of others.*" He also says: "That this second form of the struggle should all but have escaped the notice of evolutionists is the more unaccountable since it arises, like the first, out of the fundamental functions of living organisms. Seldom has there been an instance on so large a scale of a biological error corrupting a whole philosophy.

"Instead of a cold, hard condition of life, the realm

of animal development is marked by a warm and unselfish concern for the welfare of others. . . . Two functions are discharged by all living things. The first is nutrition, the second is reproduction. The first is the basis of the struggle for life, the second is the basis for the struggle for the life of others. They are involved in the nature of protoplasm itself."

Then Drummond, in a very beautiful statement, describes *the wonderful fact of motherhood* in nature. He asserts: "Without some rudimentary maternal solicitude for the egg in the humblest forms of life, or for the young among the higher forms, the living world would suffer, and would cease."

NATURAL SELECTION DISPROVED BY MENDEL'S LAW

A most revolutionary discovery came with thorough testing of Mendel's law. Gregor Mendel, an Austrian, after many years of experimentation, wrote his first paper on *Experiments in Plant Hybridisation,* in which he proved that offspring may show the character possessed by either parent; but that *it cannot develop any characters whatever which were not manifest or latent in the ancestry.* He emphasised amazing possibilities in varieties, but proved that after a certain extent of variations had been reached, the transmission ceases, and recurrence to type is observed.

Two outstanding students of Mendel's law and his experiments are Dr. William Bateson, Professor of Biology in Cambridge University, and Dr. Thomas Hunt Morgan, Professor of Experimental Zoology in Columbia University. We quoted both of these scientists, and mentioned the books published by them, in the beginning of this chapter. As a result of their findings they declare the theory of natural selection no longer tenable, as do other leading scientists.

Prof. Bateson gives a technical discussion of Mendel's work, a biography of Mendel, and two of his principal papers. Prof. Bateson says: "The conception of evolution as proceeding through the gradual transmission of masses of individuals by the accumulation of impalpable changes is one that the study of genetics shows immediately to be false. Once for all that burden, *so gratuitously undertaken in ignorance of genetic philosophy*, by the evolutionists of the last century, may be cast into oblivion. That the control of variations is guided ever so little in response to the needs of adaptation, *there is not the slightest sign.*"

Last December (1921), Prof. Bateson gave an address at Toronto before the American Association for the Advancement of Science, in which he said: "It is impossible for scientists longer to agree with Darwin's theory of the origin of species. Varieties of many kinds we daily witness, but *no origin of species*. Thus the progress of science is destroying much that till lately passed for Gospel." Even Haeckel admitted that "no such thing as well-defined species in the dogmatic sense of the schools has ever appeared."

BATESON'S POSITION AROUSES DISCUSSION

Prof. Bateson's address caused a decided stir. In *Science* for February 24th, 1922, Prof. Henry Fairfield Osborn, of Columbia, hastened to the defence of the generally accepted views. He declared that the theory of natural selection has much to be said in its favour. But Prof. Osborn *does not deny Bateson's assertion* that we do not witness any origin of species by natural selection. In fact it is evident that Prof. Osborn does not feel very sure of his ground, for he says: "If Bateson's opinion is generally accepted as a fact, or demonstrated truth, the way is open to search the causes of evolution along other lines of inquiry."

We have quoted Hugh Miller, Huxley, Haeckel, Mendel, Bateson, Prof. J. Arthur Thompson, and even Darwin himself, to the effect that men have never observed new species appearing by the evolutionary process. Yet Prof. John M. Coulter, of the University of Chicago, in a recent article in the *Christian Century,* insists that many species have been observed to originate by evolution, and are now going through the process. *It is strange that the great scientists mentioned above know nothing of these instances mentioned by Prof. Coulter.*

In his article Prof. Coulter stresses the fact that many "intergrades" are to be noted; but intergrades are always varieties, not species. This suggests that Prof. Coulter must be numbered among those scientists who incline to call varieties species, which are not true species according to the tests mentioned above, as fixed by leading scientists.

On this point Prof. Bateson says: "We may be certain that numbers of the 'recognised species,' if subjected to breeding tests, would immediately be proved to be *only analytical varieties.*" In accord with this view Dr. David Starr Jordan, the leading authority on fishes in America, says, in his *Science Sketches:* "In our fresh water fishes each species has been described 'new' three to four times, on account of minor variations, real or supposed."

Another high authority makes the same statement. Prof. H. D. Scott, of Edinburgh, published in *Science* for September, 1921, an article, in which he said: "It has long been evident that *all those ideas of evolution in which the older generation of scientists grew up* have been disturbed, or indeed transformed, since the rediscovery of Mendel's work, and the consequent development of the new science of genetics. The small variations, on which the natural selectionist relied so much, have proved for the most part to be merely fluctuations, oscillations about a mean, and, therefore *incapable of giving rise to permanent types*."

UNIFORMITY OF FOSSIL RECORDS DISPROVED

Until recently a general idea obtained that Geology furnished us the record of a succession of different types of life on this globe, revealed in fossil remains, presenting a well-defined order from the lower to the higher. The theory of evolution was largely built on this assumption. But the assumption was wrong. We have quoted Hugh Miller's positive proof to this effect. In spite of such evidence as Miller presented, scientists persisted in promulgating the theory.

Prof. Huxley, in his *Discourses,* candidly says: "In the present condition of our knowledge and of our methods, one verdict—not proven and not provable— must be recorded against all grand hypotheses of the palæontologist respecting the general succession of life on the globe."

The ordinary textbooks generally teach that in the

oldest rocks the simplest forms of fossils are found, while the more complex and highly developed forms are found in the later strata. We have seen that Hugh Miller proved this to be untrue. It is now proved that it is impossible to arrive at any certainty regarding a chronological order.

Prof. George McCready Price, of the University of Southern California, author of *The Fundamentals of Geology*, has spent twenty years in personal investigation of the geological formations of the Pacific coast region. He reports that in vast sections of the Northwest, especially in Montana and Alberta, he has found, "in numerous cases the usually expected conformable conditions exactly reproduced upside down." That is to say, very "old rocks" occur with just as much appearance of natural conformability on top of very "young rocks," and covering hundreds of miles, "in some sections covering five or six thousand square miles of area."

Until recently such irregular formations have been explained as being "thrust faults," meaning that a portion of the earth's crust has been pushed up on top of other portions. This view was held when only small areas were involved; but scientists generally admit that this theory cannot explain so vast an area as exists in North America.

Moreover, it is proved that *the same fossils are in all of these different formations*. Hence Prof. Price declares: "The facts in the rocks prove that the common geological distinctions as to age between fossils are unjustifiable. In any particular locality, of course, the lower rocks are older than the upper ones; that is to say, they were deposited first. But the so-called geological succession is a purely artificial classification. For instance, the use of the graded series of fossil 'horses,' is as inconclusive as an arrangement of modern dogs from the little Spaniel to the St. Bernard. Such series *are simply arrangements* of fossils found, with *no proof as to the order of their first appearance."*

NEW REVELATIONS IN RADIOACTIVITY

In 1896 Prof. Alexandre Bequerel discovered radioactivity, and a new experimental science came into existence. It derives from nature at first hand astonishing evidence of the properties of atoms previously unsuspected. Chemistry never dreamed of the changes which are spontaneously taking place in certain elements of matter.

Two of the outstanding scientists who deal with this subject are Prof. John Joly, of the University of Dublin, in his book *Radio and Geology;* and Prof. Frederick Soddy, of the University of Glasgow, in his book *Matter and Energy.* Both books are marked by clarity, and are popular in style, though technical. Prof. Joly makes the significant statement that "there are many, even among scientific readers, who are still unacquainted with the considerable body of facts which enters into the subject of radioactivity *as an influence on terrestrial history.*

A few general statements found in these books may prove helpful. A radioactive substance is one whose atoms are marked by a lack of stability, and undergo partial disruption brought about by *an initial velocity inherent in the atoms themselves.* The energy which leads to this change is not imparted from without, but is inherent in the atoms, and is constantly given off in emanations.

Science is acquainted with two families of radioactive substances. One is the Uranium family. The other is the Thorium group. The uranium family includes twelve known or inferred elements, one of which is radium. Each element is *derived from its predecessor in the series by a loss of emanations.* This process of constant diminution is accompanied by the presence of heat.

After uranium has given off certain emanations, the result is ionium. When ionium has given off certain emanations, the result is radium. After radium has given off certain emanations, the result is helium, and so on. Each element has its "period of transformation." That of radium is comparatively short-lived, being only 1,760 years.

Uranium is much more stable than radium. There is about three million times as much uranium as radium in the earth, and its life is about three million times as long. Prof. Joly says: "The efficacy of uranium as an almost eternal source of thermal energy seems to be unquestionable. We need not submit uranium ore to either chemical or physical processes. It is sufficient to take it from the rocks and place it in the calorimeter, when the constant flow of heat will be apparent."

THEORIES ABOUT THE AGE OF THE EARTH REVISED

The bearing of this discovery upon the theory of evolution is to be noted at two points. It was long supposed that the crust of the earth has been formed by cooling, and that the gradients of temperature, which had been measured for a considerable depth, served as a basis for calculating the time of the cooling process. This idea rested upon the pre-supposition that the earth was a molten ball, thrown off from the sun, and that it is molten within this cooling crust.

But Prof. Thomas C. Chamberlin, of the University of Chicago, represents a group of modern scientists who tell us the earth is solid. Radioactivity gives us a new explanation of the heat in the earth, as being the result of the constant emanation of radioactive bodies. This explains the molten mass thrown out by volcanoes, as the heat moves toward the surface along lines of least resistance.

Prof. Soddy says: "The day is gone by when the earth is regarded as simply a cooling world. It has in its known material constituents a steady source of fresh heat. Instead of growing cooler by radiation, it is regarded as steadily growing hotter in its interior. . . . At some time in the future, a world so constituted must explode, when the increasing temperature and pressure within overpower the strength of the crust."

In the light of these new discoveries scientists have realised that the age of the earth must be calculated on a new basis. Prof. Joly describes at length the considerations which enter into this calculation. He reports Lord Kelvin's estimate as being not less than twenty millions, and not more than forty millions of years, with a leaning toward the smaller figure. He also reports Prof. Solas as estimating the earth's age as twenty-six millions of years. Joly is convinced that Lord Kelvin's estimate "is not likely to be controverted." He further declares that "solar events must be comparatively short-lived," and that "the whole cosmos may have entered its present phase of existence within a period correspondingly recent."

We have quoted Darwin and others to the effect that the evolution theory requires many ages for realisation, especially since we have no evidence of any evolution taking place within known fossil history. Sir George Darwin, in his Presidential address at Capetown, in 1905, said: "It does not seem extravagant to suppose that five hundred to one thousand million years may have elapsed since the birth of the moon." Other scientists have held to similar figures.

Radioactivity, in causing new calculations which prove the earth to be short-lived, strikes a serious blow at the theory in general, since we now know the long ages necessary to evolution did not exist.

Prof. Soddy says: "The difficulty with the elder physicists was to allow geologists sufficient periods of time for the processes they studied. That was before these processes of radioactivity were known, in which the energy involved is a quarter of a million or more times greater than in any previously known process."

THE PRESENT COSMIC PROCESS HAD A BEGINNING IN TIME

Evolutionists have argued that the present process in nature is a continuation of an endless process. Radioactivity has proved that this assumption is false. The spectroscope proves that radioactive elements in the sun and stars are the same as those in the earth, and subject to the same law of radiation. We have noted that a constant emanation is going on, as a result of which each radioactive substance is being steadily diminished. This process is going on throughout the physical universe.

Moreover, there is no evidence that any replenishing of the loss sustained by each radioactive body is taking place. Prof. Soddy gives us a homely illustration in the case of the coal supply in the earth, which is steadily being diminished in emanations of heat, with no indication anywhere of a provision to replenish the loss.

Had the physical universe, in its present process, existed from endless ages, it would have been utterly exhausted by giving off thermal energy long since. Our supply of solar energy would have been gone ages ago. This means that the physical universe in its present form is *a stupendous clock that is running down.* It also means that the present cosmic process *began at a definite point of time,* and that it is *a different process from that which may have gone before.* Hence the evolutionists are wrong again. Moreover, it must be evident that a direct creative agency started this present process according to a new and specific plan.

SPONTANEOUS GENERATION OF LIFE AN UNSCIENTIFIC ASSUMPTION

Our findings point to the fact that science demands creative intelligence and power at certain great epochs in the creative process, in order to an adequate explanation of the same. Evolutionists claimed that life appeared by spontaneous generation. That is to say, one day dead dirt suddenly clothed itself with the attribute of life. But no scientist ever claimed that any one ever discovered an instance of such spontaneous generation. *All experiments repudiate the assumption.*

Prof. Tyndall made very extensive investigations in this field of research. He said: "I share with Virchow the opinion that the theory of evolution in its complete form involves the assumption that at some period of the earth's history there occurred what would be called spontaneous generation of life; but I also agree with him that the proofs of it are still wanting. I also hold with Virchow that the failures to discover such spontaneous generation of life have been so lamentable that the doctrine is utterly discredited."

The only alternative is the presence of a direct creative power. Alfred Russell Wallace declared: "The very first vegetable cell must have possessed altogether new powers. Here we have an indication of a new power at work." In his first edition of *The Origin of Species,* Mr. Darwin held this view. He said: "There is a grandeur in this view of life, with its several powers, having been originally breathed by the Creator into a few forms, or into one."

CONSCIOUSNESS ALSO INDICATES DIRECT CREATION

The next epochal moment in the process of nature was when consciousness came into existence in animal life. Leading scientists agree that the then-existing nature could no more produce consciousness of itself than it could produce life. Wallace said: "The advance from the vegetable to the animal kingdom is completely beyond all possibility of explanation by matter, its laws and forces. It is the introduction of sensation or consciousness, constituting the fundamental distinction between the animal and vegetable kingdoms."

THE HUMAN MIND DIFFERENT IN KIND FROM THAT OF THE LOWER ANIMALS

The point of importance is whether the difference is one of degree simply, or one of kind. Darwin naturally insisted that it is only a difference in degree. Otherwise he must surrender his theory. But other scientists deny the assumption, insisting that man possesses a distinct endowment, unique in kind.

Prof. Romanes, in his *Mental Evolution of Animals,* asks the question: "Wherein does the distinction truly exist?" His answer is: "It consists in the power which the human mind displays of objectifying ideas, or setting one state of mind before another state, and contemplating the relation between them. This is the power to think, by introspective reflection, in the light of self-consciousness. We have no evidence to show that the animal is capable of thus objectifying its own ideas. Indeed I will go further and affirm that we have the best evidence to prove that no animal can possibly attain to these excellences of subjective life."

Prof. Lloyd Morgan, the Naturalist, in his *Animal Life and Intelligence,* says: "I do think that we have in the introduction of the analytical faculty so definite and marked a new departure that we should emphasise it by saying that the faculty of perception, in its various specific grades, *differs generically* from the faculty of conception. And, believing as I do, that conception is beyond the power of my favourite and clever dog, I am forced to believe that *his mind is different generically*

412

from my own." Here again the inevitable alternative is direct creative agency.

MAN IN A CLASS BY HIMSELF

Thus far we have not quoted at length from Alfred Russell Wallace, for the special reason that we desire to emphasise his position respecting man's unique place in creation, demanding special creative plan and power. Wallace accepted natural selection as probably obtaining in nature below man; but refused to consider it as a reasonable explanation of the nature of man. He insisted that science demands special creation in the case of man.

In his books *Darwinism* and *On Natural Selection,* Wallace sets forth his convictions "that man's entire nature and all his faculties, whether moral, intellectual or spiritual, have not been derived from their rudiments in the lower animals in the same manner and by the action of the same general laws as his physical structure has been derived. . . . The inference I would draw from this class of phenomena is that a *superior intelligence guided the development of man in a definite direction and for a definite purpose,* just as man has guided the development of many vegetable and animal forms."

Had Wallace known, as we know now, that natural selection did not obtain in the origin of man, nor in the origin of lower animals, his general attitude makes us confident that he would have been one of the first to welcome the findings of Mendel's law. He indicates several points in his contention.

The lower animals have *a hairy covering.* Man in

a savage state needed this covering, since it would be useful for his protection. Evolution demands that any useful part be retained and improved. But hair is absent from most of man's body. It is thickest on the backs of animals, but usually absent on the backs of men. No theory of evolution can explain this loss.

Again the Quadrumana *go on "all fours,"* horizontally or in a stooping position, while man walks upright. The brute has powerful muscles in the back of the neck to carry its head in this position, while man has no such muscles. Darwin held that man had been evolved from an "ape-like progenitor"; but these animals use the big toe as a thumb, while man has no such prehensile use of his big toe. Moreover, the great superiority of the hand of man over the forefoot of any lower animal indicates something more than a natural process of evolution to explain its origin.

The brain of man specially excites the wonder of Wallace. Evolution would give man a brain capacity slightly superior to that of the next lower animals, since this would have been sufficient to maintain superiority on man's part. But man has a much greater brain capacity than is required to maintain this superiority. Moreover, some of the largest brains have been found in savage men, and as we trace back to pre-historic man, there is no material diminution of the brain case.

Man's *unique power of language* is another distinctive endowment. Max Mueller says: "Between the language of animals and that of man there is no natural bridge. Human language such as we possess requires a faculty of which no trace has ever been discovered in lower animals. Rational language is traced back to roots, and every root is the sign of a general conception or abstract idea, of which the animal is incapable."

Prof. Alfred Fairhurst, in *Organic Evolution Considered*, makes the keen comment that "we do not know that the size of the brain is in any way dependent on language. Ideas precede words, and faculties precede ideas. Ideas invent words. If the ape had ideas, he would invent language to express them, especially if he is the ancestor of man, who has invented a great multitude of languages."

It is generally known that modern evolution no longer follows Darwin's idea that man is descended from the ape, for they can no longer hold his view scientifically in view of these facts. Therefore, in order that they may not be compelled to surrender the theory altogether, they have invented the idea that both man and the ape descended from a common ancestor, *of which there is not the slightest trace known in the fossil history of animal life.* Could anything be more unscientific? Is this intellectual honesty?

Having invented this gratuitous assumption, they go merrily along, as if they had a right to call themselves scientific in their methods! Prof. Dana, in his *Manual of Geology,* says: "Man's origin has thus far no scientific explanation from science. The great size of his brain, his eminent intellectual and moral qualities, his voice and speech, give him his sole title to the position at the head of the kingdom of life."

Most significant of all is the fact that savages in all parts of the globe not only have a brain capacity far greater than is demanded by their mode of living, but are quite *capable of a rapid education to the point of leaving their savage life behind in a remarkably*

415

short time. If evolution were true, this would be impossible, for the powers of savage men would not have persisted if not used. Moreover, savage men have proved to *possess moral and religious capacities* which enable them, in less than fifty years, to accept the truth of the Christian religion and respond to its constraints upon them to turn from their old life and develop genuine Christian character.

Concerning this extraordinary capacity of savage men, Prof. Fairhurst says: "The fact of such great and sudden changes produced in the lives of the most degraded savages shows the infinite gulf between them and the highest brutes. The more degraded man is shown to be in his savage condition, the more wonderful becomes the contrast between him and the highest animals, when he has the opportunity of civilisation. Evolution, instead of gaining, loses much by hunting up degraded savages, for the lowest tribes have vastly more capacity than evolution calls for, or can explain.

Darwin, in his *Descent of Man,* concedes the unique character of man's moral nature. He says: "The moral sense perhaps affords the best and highest distinction between man and the lower animals. *Man alone can with certainty be ranked as a moral being.* He alone is capable of comparing his past and future actions or motives, and of approving or disapproving of them. We have no reason to suppose that any of the lower animals have this capacity."

TEN FINDINGS AGAINST EVOLUTION

Let us sum up our findings as to the opinions of leading scientists regarding the evolution theory.

First, The processes of the physical universe have not always been the same, as evolutionists claimed; but the present cosmic system is a stupendous clock that is running down, proving that at a given time it had its beginning, and is something different from what it was before; indicating direct creative plan and power.

Second, The age of the earth, according to the latest calculations, in the light of the new science of radio-activity, is about thirty millions of years. Therefore the "long-past geological ages" as assumed by Darwin as necessary to allow the transmission of imperceptible persistent characters, necessary to the theory of evolution, did not exist.

Third, Palæontology can no longer be considered an accurate witness as to the assumed regularity of the order of fossil remains from lower and simpler forms to higher and more elaborate, for scientific investigation has proved that such order did not and does not exist. On the other hand, all the available fossil history does not reveal one instance of the evolution of species.

Fourth, Every known fact, as the result of careful investigation, disallows the theory of spontaneous generation of life, as the theory of evolution requires. Again the only alternative is direct creative agency.

Fifth, The struggle for life, as emphasised by Darwin, was not the greatest fact to be stressed; for the struggle for the life of others is of equal, if not greater, importance. Nutrition is accompanied by reproduction in order to continued life; and motherhood makes a unique contribution to all intelligent life. This fact points to unselfish concern for created things, suggesting direct creative agency.

Sixth, Haeckel's theory of genetics in embryology, involving the so-called recapitulation theory, is no longer accepted by those scientists who have tested it thoroughly; but is declared positively disproved. Thus one of the most impressive arguments for evolution must be discarded.

Seventh, Variations, while appearing in great numbers and striking varieties, do not persist through long ages, but recur to type, as proved by Mendel's law; and the theory of natural selection is repudiated by the latest scientists who have been original investigators.

Eighth, Consciousness did not come just by "the jostling of atoms together," for the gulf between the unconscious matter and the conscious animal has never been bridged from below. The alternative explanation of this new capacity is direct creative plan and power.

Ninth, The human mind is not simply greater in degree than that of the lower animals, but is generically different in kind. This cannot be harmonised with the theory of evolution, and points to direct creative power.

Tenth, The distinctive characteristics and capacities of man, especially his moral and religious endowments, are impossible of explanation by the evolution theory, so that science demands recognition of direct creative purpose and power in explanation of man's origin and progress.

WHAT MUST BE THE ATTITUDE OF SCIENCE TOWARD EVOLUTION?

When Herbert Spencer realised that so many scientists were taking ground against various points of the evolution theory, he felt bound to defend the idea that acquired characters persist through ages of transmission. In the *Contemporary Review*, February-March, 1893, he wrote: "Close contemplation of the facts impresses me more strongly than ever with two alternatives—either there has been inheritance of acquired characters, or *there has been no evolution.*"

This candid admission, in the face of the present proof that acquired characters do not persist, reminds us of Prof. Bateson's statement that if Darwin had known the facts now proved regarding Mendel's law, he would never have written his books. And we are compelled to believe that if Spencer were now alive, he would stand by his alternative and declare that "there has been no evolution."

But modern evolutionists do not accept Spencer's alternative. Both Bateson and Scott assert their loyalty to the evolution theory, in spite of the fact that they repudiate the old assumptions that biology furnishes any proof of it. They tell us they fix their faith in evolution because of the findings in palæontology. Yet we have shown that palæontology is no longer competent to furnish any facts to justify their faith.

All along we have been indicating that there is an alternative explanation of the origin of created things. Prof. Haeckel had this alternative in mind when he declared that, rather than agree with Weissmann and Wallace, in denying the inheritance of acquired characters, "it would be better to accept a mysterious creation of all the species as described in the Mosaic account." This is exactly what Hugh Miller insisted upon, as did Wallace, Mendel, Agassiz, Virchow and other leading scientists.

Prof. Dana, of Yale, in his little book, *Genesis and Science,* points out that the order of creation is exactly that indicated in Genesis. Cosmic light, from the earth's point of view, is followed by the separation of the planet from the firmament, the division of land and water, and the beginning of vegetation during the carboniferous period. Then sun and moon and stars appear, in the perspective of their values to the earth. The simpler forms of animal life follow, in accord with the record of science, followed by more complex forms, ending with the account of the special creation of man.

Prof. Dana declares that when we compare this Genesis account with the puerile cosmogonies of other sacred writings, its sublimity of statement, and its substantial harmony with the findings of modern science convinced him that no man could have written it long centuries ago, without divine inspiration.

The statement that Genesis teaches creative days of twenty-four hours each is disproved by the text itself. In Gen. 2:4, we read: "These are the generations of the heavens and the earth when they were created *in the day* that Jehovah God made the earth and the heavens." Here we have clear room for geologic creative days, as the term covers the whole period of creation.

Why should people who claim to be scientific hesitate to accept these findings of the leading scientists disproving the evolution theory? Various answers will probably be given to that question. Surely we have the right to demand the evidence of intellectual honesty on the part of those who would be our scientific leaders. We cannot do less than demand facts which will offset the findings which disprove the theory at so many points, if men refuse to follow the logical conclusion of those findings. No facts have been yet produced. Science demands that preconceptions must be cast aside, no matter how long or how strongly they may have been held, when facts make them unjustifiable.

We have had no antecedent prejudice against this or any other theory about the processes in nature, so long as they recognise the hand of the Creator, if only said theories are validated by facts. But since we find all the facts thus far presented by original investigators in the several fields of research involved, proving that the organic evolution of species can be no longer assumed; we refuse to follow the advocates of that theory, and must look elsewhere for light upon the problems of creation and life.

Chapter VIII

THE SPIRITUAL INTERPRETATION OF THE UNIVERSE

In considering all questions of origins, we must keep in mind our conscious experiences as furnishing the basis for any intelligent convictions regarding the explanation of our presence on the earth, and the kind of beings we are. This must also be the basis for our convictions as to the world at large. Let us apply this principle to our interpretation of creation.

We make things. In doing so, we are conscious of exercising the desire and the will to make the particular thing created. We study our constitution, and realise how marvellously the law of adaptation is revealed in its many parts. We know that when we exercise our will power, our bodies respond instantly in obedience to our wills. I decide to lift up my hand, and instantly my hand comes up. In a former chapter we referred to the fact that Prof. Romanes came to appreciate the truth that this self-conscious, intelligent, volitional being is of the nature of spirit.

NATURE REVEALS INTELLIGENT WILL IN ACTION

Alfred Russell Wallace, in his *Natural Selection,* says: "Force is the product of mind. All force is probably will-force. If will is anything, it is a power that directs the action of forces stored up in the body, and

422

it is not conceivable that this direction can take place without the exercise of some force in some part of the organism. If, therefore, we have traced one force, however minute, to its origin in our will, while we have no knowledge of any other primary cause of force, it does not seem an improbable conclusion that all force may be will-force; and thus that the whole universe is not merely dependent on, but actually is, the will of higher intelligences, or of one Supreme Intelligence."

As an accompaniment to the evolution propaganda, a great wave of naturalism swept over the thinking world. The extreme materialist developed the theory that matter is the source of everything, and that all manifestations of intelligence and moral appreciation are just the results of an age-long evolution out of matter, by *accidental* developments, and by the action of resident forces which act and react with mechanical uniformity.

Materialism would declare that the action involved in the raising of my hand was due to the energy inherent in the muscles, involving a mechanical necessity. The sufficient answer to that is the fact that I can determine *when* I will raise my hand. I decide to wait five minutes, or an hour, and then the moment I exercise my will, my hand comes up. Thus I prove that I have a free will, and our study will prove that *freedom is the outstanding feature* of the human will.

INTELLIGENT DESIGN EVIDENT IN NATURE

Immanuel Kant was sympathetic with the materialistic theory for a while; but he abandoned it as being untenable. His mature judgment is stated thus: "It is impossible to contemplate the fabric of the world without recognising the certain manifestation of the hand of God in the perfection of its correlations. Reason, when once it has considered and admired so much beauty and so much perfection, feels a just indignation at the dauntless folly which dares to ascribe all this to chance and happy accident. It must be that the highest wisdom conceived the plan, and infinite power carried it into execution. All things which set forth reciprocal harmonies in nature must be bound together in a single Existence on which they collectively depend."

Lord Kelvin, in his Chancellor's address, in April, 1903, said: "Science positively affirms creative power. It is not in dead matter that we live and move and have our being, but in the creating and directing Power which science compels us to accept as an article of belief. We are absolutely forced by science to believe in an influence other than physical, or dynamical, or electrical forces. There is nothing between scientific belief in a Creative Power and the acceptance of the theory of a fortuitous concourse of atoms. Modern scientific men are in agreement in condemning the latter as utterly absurd in respect to the coming into existence, or the growth, or the continuation of molecular combinations presented in the bodies of living things."

"Forty years ago I asked Liebig, walking in the country, if he believed that the grass and flowers that we saw grew by mere chemical forces. He answered: 'No, no more than I could believe that a book on Botany, describing them, could grow by mere chemical forces.' Every action of free will is a miracle to physical and chemical and mathematical science. Do not be afraid to be free thinkers. If you think strongly enough, you will be forced by science to belief in God, which is the foundation of all religion."

Prof. James H. Snowden, of Pittsburgh, in his book, *The Personality of God,* says: "There is not a particle of unreason or mental absurdity in the whole universe. The world is found to be a mental construction that reveals the presence and working of a Mind as certainly as a book reveals to us the mind of the author."

BENEVOLENCE EVIDENT IN CREATIVE THOUGHT

Moreover a benevolent purpose is manifest in nature. Every law in nature is a good law, because its obedience brings only and always blessings. That is to say, love is manifest in all law. A penalty always follows the violation of law. This is necessary to the defence of the values preserved by the law. The law of purity is only maintained by the love of purity, and that love must be a burning flame against impurity, in defending its priceless values to mankind. Without the penalty, the value of the law would cease.

Hence it is that, when law is disobeyed, the ills and ails of life appear. Very often men are not willing to follow the leading of the law as a guide to the way of blessing. *The loss is always registered in the character* of the individual or the community or the nation. When we disobey the law of purity, we impair our purity, and the penalty is the tragic loss of this priceless value in character. There is nothing arbitrary about this on the part of the Ruler of the moral government of the universe, as some people imagine. *It is the result of our own choice,* when we refuse to follow the law as the guide to the way of blessing.

It is tremendously important to realise that *it is man's wilful disobedience of law,* which plunges individuals and nations into all sorts of tragedies and losses in the realm of human values. The God of love has pointed the way of blessing in his law. He will not force the free will of men. It is *pitiful short-sightedness on any one's part to deny the love of God* because of the ills which men bring upon themselves through disobedience. It is equally short-sighted to suppose that the Law Giver could ignore man's disobedience. Such a program would mean the loss of all value in law.

The benevolence of the Creator is evident in many facts in nature. One is *the atmosphere,* which is necessary to life. Wallace, in his book, *Man's Place in the Universe,* holds that our earth is the only habitable sphere in all the known worlds, as we now know the conditions of life, for it alone has an atmosphere like ours. Its most striking feature is the abundance of oxygen, which is necessary to life. Although oxygen combines most easily with many substances, its great surplus is an indescribable boon to man.

The *abundance of water* is another unique feature, so essential to life. It is also essential to the soil, in the growth of vegetation and food for animals and men. It also has multiform uses in all sorts of hygienic, mechanical and commercial activities. Then we see in nature an almost unbelievable co-ordination of conditions in the environment, and the adaptability of the creatures to these conditions, as in provision for food, conditions of climate, and many other adaptations leading to the blessings of life.

Moreover, there is evidence of *redemptive power* in nature. If you cut a tree, immediately certain healing potencies are released which seek to heal the hurt. We have noted the fact of motherhood and its altruistic value in all life. Parenthood is marked by eager effort to nurture and guide offspring in the ways of safety and prosperity. Among men, as this grade of noblest culture improves, this parent love is ever ready to forgive, to restore and to assist the child to retrieve the past failure, and make good for the future. All this is an evidence of the self-revelation of the nature of the Creator. Because of manifest benevolence in nature, men have called the Creator, the Good. We have shortened this to God.

PSYCHOLOGY TEACHES THE SELF-REVELATION OF SPIRIT

Again we turn to human experience to realise the very important fact, taught by the science of psychology, namely, the self-revelation of spirit. In a cradle you see a body, but perceive no motion or sound. You conclude it to be a dead body, since there is no evidence of life. But if you perceive motion and sound, you infer that there is a little animal in the cradle. As it gives evidence of intelligence, people say it is a bright child. Or if it should reveal mental aberration, we are compelled to say it is an idiot. Every conclusion is *compelled by the character of the self-revelation* made by the child.

Whatever comes out in the self-revelation determines

our conviction regarding the spirit's presence and character and ability in every individual life. Though the spirit is invisible and intangible to the physical senses, he maintains contacts and communions with other intelligent spirits by means of the physical media at hand, according to the laws of control of these media, and the laws of spiritual fellowship.

Exactly thus do we behold the facts in the physical universe which compel the conviction as to the presence and power of a spirit-being, whose nature is like our own, revealing to us thought and purpose, as clearly as we reveal thought and purpose and power to each other day by day. Whatever has come out in nature, on the positive and constructive side of all law, is the *revelation of the Creator* in so far.

THE PERSONALITY OF GOD

We have already anticipated the direct evidence which compels recognition of the personality of the Ruler of the universe. Prof. Snowden points out that personality involves the presence of intellect, feeling and will. We have noted abounding evidence of design, benevolence and power in God, which justifies the statement of Prof. Snowden that "the universe manifests itself to us in terms of personality." The same kind of evidence that proves human personalities compels recognition of the personality of God.

Modern thought is giving much attention to this subject, for there is a revival of the pantheistic philosophy which claims that God is a principle, and not a personality. The advocates of this view seem to feel driven to it, in the fear that if they admit the personality of God, they must think of him as a subject to limitation, which they cannot admit concerning a supreme being. They seem to confuse the nature of spirit-personality with the limitations which mark the individual personage.

One of the ablest discussions of certain fundamentals of the Christian religion is by Principal P. T. Forsyth, in his book, *The Person and Place of Jesus Christ.* In discussing the subject of the limitation of personality, he says: "Personality is not limitation, nor the negation of limitation, but the surmounting of it. Determination here is not negation, but power, for it is self-determination. Mere individuality may be defined by limitations, but personality is expressed within them by transcending, overflowing and utilising them. The individual is an area, but the person is a centre of power."

GOD'S INFINITE ABILITY TO NOTE DETAILS

This is another fact to emphasise. Science teaches us of an inconceivable infinity of thought in creation which reaches past the atom of hydrogen to the electron. Hence it is quite consistent with this fact to hold to the conviction that God is able to carry in his thought an infinite number of his creatures, with an intimacy of affectionate interest, even as a mother is interested in each of her children, no matter how many there may be.

This is only one of the many capacities of God which far transcend man's finite limitations. But this one should be specially magnified in our appreciation, for it is basic to the adequate realisation of our personal

429

fellowship with God as his redeemed children, in each of whom he is definitely interested. No least thing is too small for his concern, because everything counts as a factor in the building of character, and character is the most important thing in the world to God.

THE FATHERHOOD OF GOD

Since all that appears in nature, in harmony with the laws of God, points to his character as Creator, we continue to seek for qualities in his character not already noted. At the crown of nature we find human nature, and in the realm of human life we find fatherhood and motherhood. This can only mean that we must recognise fatherhood in God, which must include the values of motherhood. Fatherhood in the creation indicates Fatherhood in the Creator. Therefore the scientific belief in God will include the appreciation of his Fatherhood. All the noblest qualities of earthly parents must be attributed to God, who exercises them without limitations.

This immanence of God in all the creation, and therefore in all of our life, must be emphasised because of its immense importance. In our limited experience we know how the human spirit reaches far beyond the point of his contact with physical media, communicating with other spirits thousands of miles away. At the same time we know that man has a basis of contact with the outer world, in his mode of subsistence, at the brain, which is his throne of power. Just so we may scientifically think of God as maintaining a throne of power, while he is everywhere present in his universe of worlds.

SCIENTIFIC CHRISTIAN THINKING

QUESTIONS ABOUT A FIRST CAUSE

Another question which keeps pushing into the realm of discussion is the inquiry about the proper idea to hold regarding a first cause. Manifestly we cannot rest in the old statement often made that every effect must have a cause, for that would compel us to demand that God also have a cause.

Finite man cannot prove a first cause. In considering what we may and must believe on this subject, the essential fact for us to emphasise is that, since life can only come from antecedent life, as we see its appearance, the Creator is a living being. Moreover, we have already considered the evidence to compel us to believe that he is an intelligent, volitional spirit.

We may hold one of two views about a first cause of all created things. First, we may insist that God had a cause which explains his existence and character. That will compel us to hold that the being who is the cause of God also had a cause. That is to say, we would hold to the idea of an endless chain of causes, which involves the necessity of believing in the eternal existence of spirit-beings endowed with life.

The second view is that God is possessed of eternal existence. This is more consistent with our appreciation of the universe than to try to think of an endless chain of living beings, which is only another way of thinking of eternal life. The universe indicates a Ruler who is sufficient unto himself, maintaining his creation with poise and power, according to his plan. We recall the statement of Prof. Romanes that science demands belief in one God, and only one, because of the uniformity of law and the solidarity of the universe.

431

SPIRITUAL INTERPRETATION

Dr. Richard L. Swain, in his book, *What and Where is God?*, tells us of a man who insisted that he could not believe in God, because he could not believe in anything that had no beginning and no ending. Then one day he suddenly realised that he must believe in space as having no beginning and no ending. This led him to the satisfactory conclusion that he could also believe in God as having eternal life. Here we have another illustration of the way we find very practical points of view in our daily experiences which involve the reasonableness of our intelligent faith in so many things that we cannot fully comprehend, but do clearly apprehend as actual experiences in our daily life.

FACTS INVOLVED IN MAN'S MORAL NATURE

Man's moral nature points to the moral nature of God, in harmony with our appreciations of the self-revelation of the Creator. We have already emphasised the fact that man is under law at every point of his being, physical, mental and moral. The realm of law indicates righteousness and justice, since obedience brings blessings, while disobedience brings penalties. This is only recognising a system of rewards and punishments in the processes of nature, including human nature.

Man is also conscious of an ethical sensibility. We give the name of *conscience* to our moral sense, which constrains us to do what we believe to be right, and restrains us from doing what we believe to be wrong. Every collective group of men lives under some form of government based on the recognition of moral appreciations and conscious individual moral responsibility. This consciousness of man's moral responsibility is the basic fact in our moral nature, for it is the responsibility of the creature to the Creator.

In the nature of things, when God made it possible for us to attain to a life of blessings, this sense of responsibility challenges us to make the best and most of our lives. It is this fact which gives meaning to religion. The word itself means "binding back," and points to this moral obligation as fundamental in man's relation to God. While we must magnify the Fatherhood of God, we must emphasise as earnestly that the moral government of God is necessary to our realisation of the meaning of life and destiny.

THE SENSE OF SIN

Man recognises his moral failure. The term "sin" is now in common usage to signify this moral failure. Evolution theories have tended to dull the sense of sin, for they led men to think of their failures as indicating the remaining weaknesses and instincts of the lower animals, suggesting ignorance and constitutional limitations, for which we are not responsible; rather than the wilful disobedience of God's law by us, who know in our hearts that we choose to disobey, and are responsible for our conduct.

Prof. Walter Rauschenbusch, in his book, *A Theology for the Social Gospel*, discusses this subject of sin, in the light of modern thinking, in a very vigorous way. He says: "Religion wants wholeness of life. We need a rounded system of doctrine large enough to take in all our spiritual interests. . . . The social gospel

433

calls for an expansion in the scope of salvation and for more religious dynamic to do the work of God. It requires more faith, and not less. It is able to create a more searching sense of sin, and to preach repentance to the respectable and mighty who have ridden humanity to the mouth of hell. We are becoming more sensitive about collective sins, in which we are involved. The social gospel is concerned with the eradication of sin and the fulfilment of the mission of redemption."

The Rochester professor declares that "the Christian consciousness of sin is the basis of all doctrine about sin. A serious and humble sense of sinfulness is part of a religious view of life. . . . When a man is within the presence and consciousness of God, he sees himself in the most searching light and in eternal connections. To lack the consciousness of sin is a symptom of moral immaturity, or of an effort to keep the shutters down and the light out. The most highly developed individuals, who have the power of interpreting life for others, and who have the clearest realisation of possible perfection, and the keenest hunger for righteousness, also commonly have the most poignant sense of their own shortcomings."

MAN IS INSUFFICIENT UNTO HIMSELF

We have quoted from Prof. Rauschenbusch for two reasons; First, because no scientific study of human nature and human needs can escape a candid recognition of the fact of human sin, and its fatality in the realm of human values, unless it be overcome. Second, because we must face, with equal candour, man's individual responsibility to God, and to his fellow-men, for his moral failures. For example, no honest man will deny that an unfaithful father or mother is responsible to God and to the children. Only as one is ready, in utter honesty, to face this responsibility and accept it, there is no possibility for him to be any better. This is bed rock, and here we must stand to enter the way of a new and better life.

Moreover, we must frankly face the testimony of history that man has never made promising progress in trying to improve his moral conduct, in his own strength. The Greeks climbed to as noble heights, two thousand years ago, in athletics, in æsthetics, in literary culture, in philosophical skill, as man has ever attained since; but they failed for the lack of moral fibre, without which there can never be stability of civilisation or permanency of greatness.

Alfred Russell Wallace, in his *Social Environment and Moral Progress*, declares that, while man has made remarkable progress in certain directions of material advancement and intellectual attainment; yet from the standpoint of morals man is to-day elevated very little above the earliest conditions *that history records.* Guesses about prehistoric man are not scientific.

Let us realise that we must take into our purview the history of the whole race, as it exists to-day on all continents, and not simply think of the favoured peoples in any part of the planet. The great question is not how far the modern self-binding, self-thrashing harvester surpasses Ruth's sickle. The great question is as to how far the womanhood of to-day has surpassed the womanhood of Ruth.

SPIRITUAL INTERPRETATION

MAY WE EXPECT THAT GOD WILL PROVIDE FOR MAN'S SPIRITUAL NEEDS?

Man's spiritual needs are manifest and supreme, if he shall ever realise victory over the power of sin. Is it reasonable to believe that God will provide for these highest needs of men? We have noted the revelation of a redemptive love in the very laws and potencies of nature. We have noted how God has provided food exactly adapted to the needs of every creature.

One thing is certain, in the light of human history. With all the advancement of science, with all the material progress of men in matters of mechanical invention and efficiency, with all the cultivation of the human intellect, *there is not sufficient light in nature to solve the problem of the human soul in respect to the way to realise victory over sin, and the attainment of the highest spiritual values in building character.* Man's *need of light is persistent,* and every honest man recognises it.

We have noted that all other human capacities find satisfaction in the provision of the Creator. Shall this highest need fail of fullest satisfaction? Nature, science, reason, faith, hope, love, all unite in confirming the conviction that our God, as we know him in nature, will not fail to give to his needy children the adequate light of life to guide us into the way of the fullest realisation of our noblest capacities and powers.

Made in the USA
Columbia, SC
09 July 2022